How MUCH is that CURE in the WINDOW?

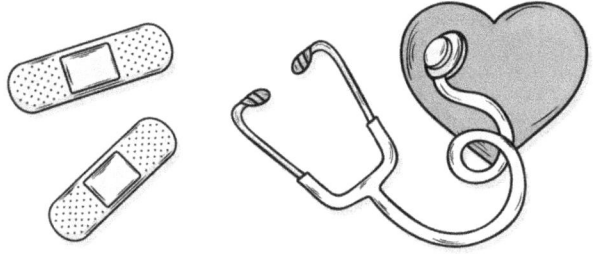

How MUCH is that CURE in the WINDOW?

Simple Math Solutions for Complicated Problems in Biology, Medicine, and Healthcare

SHAUN COMFORT, MD, MBA

Copyright © 2020 by Shaun Comfort.
All rights reserved.

No part of this book may be used or reproduced in any manner whatsoever without written permission, except in the case of brief quotations embodied in critical articles and reviews. Requests for authorization should be addressed to Shaun Comfort: stochasticmonkeypress@gmail.com.

Cover design by Ivica Jandrevic
Interior layout and design by www.writingnights.org
Book preparation by Chad Robertson
Editing by Chantal Matkin Dolan, Elayne Wells Harmer, and Diana Matkin

ISBN: 979-8-6456-5138-1
LIBRARY OF CONGRESS CATALOGING-IN-PUBLICATION DATA:
NAMES: Comfort, Shaun, MD, MBA, author
TITLE: How Much is that Cure in the Window? Simple Math Solutions for Complicated Problems in Biology, Medicine, and Healthcare/ Shaun Comfort
DESCRIPTION: Independently Published, 2020
IDENTIFIERS: ISBN 9798645651381 (Perfect bound) |
SUBJECTS: | Non-Fiction | Science | Mathematics | Healthcare |
CLASSIFICATION: Pending
LC record pending

Independently Published
Printed in the United States of America.
Printed on acid-free paper.

Although the author and publisher have made every effort to ensure that the information in this book was correct at press time, neither author nor publisher assumes any liability to any party for any loss, damage, or disruption caused by errors or omissions, whether such errors or omissions result from negligence, accident, or any other cause. Both author and publisher hereby disclaim any liability to any party. Readers should contact their attorney to obtain advice before pursuing any course of action.

Biblical references: New International Version

24   23   22   21   8   7   6   5   4   3

To my wife, thank you for all the encouragement during the writing of this. It meant the world…

I would have you realize, my lord, that men are at the mercy of circumstance, and not their master.
—Artabanus advising Xerxes before he made the disastrous decision to invade Greece (Herodotus, *The Histories*)

I find a nice long whisky and soda a great solvent of human idiocy.
—Sondelius speaking to Martin Arrowsmith (Sinclair Lewis, *Arrowsmith*)

CONTENTS

Acknowledgements xi
Chapter 1   Why? 1
Chapter 2   Fermi Estimation 6
Chapter 3   Notation and Numbers 10
Chapter 4   The Metric System, Unit Cancellation, and Constants 13
Chapter 5   From Ranges to Distributions 15
Chapter 6   Uncertainty Propagation 23
Chapter 7   How Many Jellybeans in a Jar? 27
Chapter 8   Around the World in 80 Days? 31
Chapter 9   How Many Drivers Are Texting Right Now? 35
Chapter 10   Shakespeare's Monkeys 37
Chapter 11   How Many Breaths in a Lifetime? 41
Chapter 12   How Many Heartbeats in a Lifetime? 43
Chapter 13   How Many Hours of Sleep in a Lifetime? 45
Chapter 14   What Is the Duration of "Now"? 47
Chapter 15   What's the Surface Area of a Human Being? 51
Chapter 16   What's the Volume of a Human Being? 53
Chapter 17   How Many Cells in a Human Body? 55
Chapter 18   How Much Energy Do We Use Climbing Stairs? 58
Chapter 19   What Is the Metabolic Rate of a Human Being? 62
Chapter 20   How Many Words in a Book? 65
Chapter 21   How Much Memory Storage Does a Book Require? 67
Chapter 22   How Many Books in the Library of Congress? 70
Chapter 23   What Is the Speed of Reading? 74
Chapter 24   My House Rocks! 78
Chapter 25   Struck by Lightning 81
Chapter 26   How Many Doctors Are There in the United States? 85

Chapter 27	How Many Patients Can a Doctor See?	88
Chapter 28	How Long Is the Average Wait at a Doctor's Office?	90
Chapter 29	How Many Hospital Beds Are There in the United States?	95
Chapter 30	How Long Would It Take to Inoculate the U.S. Population?	97
Chapter 31	Banging Heads Together	100
Chapter 32	The Sum of All Impacts	105
Chapter 33	Plausible Clinical Trial Enrollment	109
Chapter 34	What Is the Average U.S. Individual Income?	114
Chapter 35	What Is the Leverage of the Average U.S. Family?	117
Chapter 36	What Is the Annual Cost of U.S. Healthcare?	120
Chapter 37	What Is My Health Insurance Premium?	123
Chapter 38	What Is the Cost of a Migraine?	130
Chapter 39	What Is the Cost of Alzheimer's Care?	135
Chapter 40	Introduction to Bayesian Reasoning	140
Chapter 41	Bayes Meets Shakespeare's Monkeys: Monte Carlo Bayesian Analysis	146
Chapter 42	Is This Diagnosis True with New Data?	151
Chapter 43	Which Diagnosis Is Most Likely?	156
Chapter 44	Oh Error, Why Art Thou? The Reliability of Processes	163
Chapter 45	Why Projects (and Almost Everything Else) Run Late	169
Chapter 46	Tying Things Together: From Fermi Estimation to Entropy	176
Additional References		182
About the Author		184

TABLE OF FIGURES

Figure 1.	A log-normal distribution in log space	17
Figure 2.	The same distribution in linear space	18
Figure 3.	6σ illustration	19
Figure 4.	Skewed distributions and confidence intervals	21
Figure 5.	Cylinder volume (units^3) = πr^2 h (or l)	28
Figure 6.	Cube volume (m^3) = L^3	56
Figure 7.	Climbing stairs	59
Figure 8.	Football player momentum	101
Figure 9.	Two player system momentum	103
Figure 10.	Histogram of Bayesian estimates	149
Figure 11.	Bayesian posterior graph	162
Figure 12.	Hypothetical medication order flow	165
Figure 13.	Hypothetical parallel process flow	167
Figure 14.	Gambling project manager	170
Figure 15.	Treating completion time as binary event	171
Figure 16.	Skewed completion-time distribution	174

ACKNOWLEDGEMENTS

Before starting, I want to acknowledge that many other authors and thinkers have used the Fermi approach, Bayesian estimation, and statistical thinking for problem solving. While my focus on applying the Fermi method to healthcare and pharma is somewhat new, many of these same techniques and problems are classic, and other authors have masterfully described them. Throughout my book, I will refer to others' approaches to similar problems.

I wish to thank all those who have inspired and encouraged me to take my curious hobby to the next stage and write a book. You are too numerous to name but I must mention at least a few. First and foremost, I thank my wife, Hilary, who lovingly supported all my evening calculations and writing for so many years. My family—especially my father, mother, brother, and Uncle Jim—instilled curiosity and wonder in me from an early age. My uncle showed my brother and me how to make rockets with Vienna Sausage cans, and my father showed us the moon through a telescope. I thank my physics professors, especially Drs. Swartz, Melosh, and Sprouse, for their encouragement and humor as I went through Stony Brook; I looked up to them then, and still do. Susie, Sally, and John, my friends and mentors who made all the difference in encouraging me to take on physics and medicine—I doubted much, but you never did!

I also want to thank the many great supervisors, leaders, and colleagues I've had throughout my career in physics, medicine, and the healthcare industry. David, Cynthia, Bill, Greg, Mark, Glenn, Ellis, Wilson, and Joel were all examples of the kind of leaders I wanted to be. I have been blessed to have supportive leaders and organizations that saw potential in my development, supported my innovation and ideas, and allowed someone like me to thrive and grow.

I also want to thank my colleagues and collaborators in research and work for inspiring me to bring my best to the game and to think creatively about everything from applying statistics to safety to teaching machines to find adverse events. Cartic, Darren, Jen, Meena, Shawman, and Irene: y'all are the best, and I hope you find this work entertaining while relaxing between machine-learning projects!

To the ladies in CMD Consulting (Chantal, Elayne, and Diana), thank you for your enthusiasm, interest, and honesty in taking on this book project and for helping me turn a dream into reality. I cannot say enough good things about the work you do!

Finally, I could not have written this book without the groundwork laid by others. As the twelfth-century theologian and author John of Salisbury stated,

> Our age enjoys the benefit of the age preceding, and often knows more than it, not indeed because our intelligence outstrips theirs, but because we depend on the strength of others and on the abundant learning of our ancestors. Bernard of Chartres used to say that we are like dwarfs sitting on the shoulders of giants so that we are able to see more and further than they, not indeed by the sharpness of our own vision or the height of our bodies, but because we are lifted up on high and raised aloft by the greatness of giants.[1]

Shaun Comfort, MD, MBA
April 2020

---

[1] John of Salisbury, "Metalogicon," 257.

CHAPTER 1

WHY?

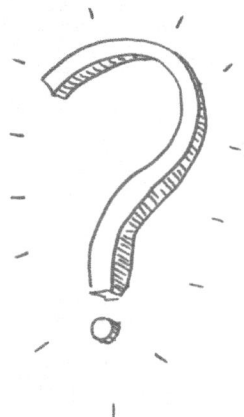

My challenge is to challenge you! Join me on a short journey through a few classic and not-so-common problems from biology, medicine, pharma, and the healthcare world (what I will coarsely aggregate into the acronym BHI or the term "bio-healthcare industry" going forward). I'll introduce you to two personal heroes of mine: Dr. Enrico Fermi, developer of the first controlled fission reaction, and Reverend Thomas Bayes, who originated "Bayes' Rule" in probability. Today, "Fermi problems" are named in honor of Professor Fermi and are commonly used as part of interview techniques for business schools and technology firms. They are also used as first approximations to solving complex problems in physics and engineering.

Seeing how someone works their way through such a problem can tell you a great deal about their ability to think. Similarly, after spending nearly 200 years as an outlier branch of probabilistic thinking, Reverend Bayes' thinking has now come to have a major impact on the world of machine learning and artificial intelligence. These gentlemen, along with others such as Ludwig Boltzmann, had much to say about the processes that generate our world, including medicine and healthcare.

I was introduced to Professor Fermi's approach during my first physics course at Stony Brook University, located about 65 miles east of New York City. I arrived at Stony Brook from Mississippi, wearing my typical outfit of cowboy boots, jeans, jacket, and a beat-up bush hat. To say I did not fit in is an understatement; I was skinny, naïve, and freezing. I had failed algebra in high school and dropped out of college to play music up and down the coast of California. For some reason, I was drawn to physics because it seemed to have the greatest promise for helping me understand "why": Why was the world the way it was? Why do so many random things happen to my family and friends? Why does time pass?

I jumped into the field and worked night and day to overcome my previous shortcomings with math, and I did reasonably well. At the time, I had no idea how physics would change my life, and how learning to reason quantitatively would lead me to a world that only Fermi could imagine. His approach became my first-line method for evaluating problems in my first career in computational electro-optics. Now, after becoming a physician and neurologist, it is still my guide to thinking about problems in healthcare, as well as a favorite pastime when I am waiting in line at an airport or hotel.

After years of enjoying this relaxing hobby, I invite you to walk through these pages with me and learn how to apply the same approaches that allowed Fermi to estimate the yield of the world's first atomic bomb with only scrap paper and a slide ruler. No instrument is needed today—just your brain, attention, and the willingness to learn. While I assume that the reader has a basic undergraduate introduction to mathematics, physics, and statistics, what I will present here is not difficult. You can easily skip any chapters that are hard to follow without losing the thread and purpose of the book.

So why write a book on quantitative, approximate thinking for healthcare? With all the emphasis and excitement about artificial intelligence and its most current manifestation known as "machine learning," why even bother with a book asking clinicians, healthcare professionals, and the pharma industry to actually sit down and estimate values? After almost a decade in computational physics and twenty years as a physician—first in practice, next as a regulator, and then in the pharmaceutical industry—I have enough experience to say that this approach is desperately needed.

My justification is that even in the twenty-first century, quickly putting pencil to paper to make a quick estimate of "what an answer might look like" has value. I've spent years presenting data and proposals for clinical trials, pharmaco-vigilance reports, project budgeting, and finance. Far too often, I find that much of the healthcare audience has an inadequate mathematical foundation to quickly evaluate feasibility, profitability, and the likelihood of success. The result is that lofty projects fail, clinical trials do not meet their enrollment goals, and health plans bust their budgets. I am not suggesting that physicians develop a working knowledge of calculus or queuing theory, but rather proposing a more modest and attainable goal: equipping interested physicians, pharmacists, scientists, and administrators with the skills to perform quick, approximate calculations that result in more successful clinic launches, fewer failed clinical trials, and potentially better outcomes for patients and practitioners. This little book is my attempt to facilitate that goal.

So now, dear reader, you may ask what you can specifically gain from joining this adventure. I think you will experience at least three benefits:

1. You will make better decisions. By seeing and working through examples of quick, approximate calculations, you can develop your ability to assess problems, propose solutions, and create

opportunities for change in your own area of the healthcare industry. This will give you a good background for making decisions and determining whether quoted numbers, rates, or costs are worth accepting or changing.

2. You will embrace imprecision and be wrong less often. Precision is expensive and precise answers take time and money to produce. By working through the problems here and in life, you will begin to see that most pursuits of precision for estimation and planning are wasted effort, inconsequential, and almost guaranteed to be wrong. A key point here is that precision is not equivalent to accuracy. In the words of Carveth Read, the British philosopher and logician: "It is better to be vaguely right than exactly wrong."[2]

3. You will develop a mindset that is difficult to fool and is comfortable with estimating answers. Be careful, though—this type of thinking will not make you popular with the "team player" and "group think" crowd that is so prevalent in business and in life. You will find that many of the great solutions and ideas for solving pressing issues in healthcare and medicine are flawed and unfeasible. However, you will likely find that your estimates will be more often correct than not, and organizations that value good ideas and analysis over politics will welcome and reward you.

I hope you will also obtain a fourth benefit: you'll have fun! You may be skeptical and feel that you could never think in this way and do these crazy word problems. However, that is not true. Starting from a failing grade in algebra and being told I was "no good at math" is about as modest a launch platform in the sciences as I can imagine. However, I did it, and so can you. Beginnings can be awkward, but so is learning

---

[2] Carveth Read, *Logic: Deductive and Inductive* (London: Simkin, Marshall, 1920), 22.

to walk. Time, effort, and curiosity will help you turn these brain-teaser problems into a habit of thought that can provide a lifetime of joy, as any crossword-puzzle enthusiast can attest. Join me as we begin!

CHAPTER 2

FERMI ESTIMATION

A famous example of "Fermi" approximation occurred in New Mexico in 1945, when Enrico Fermi and many of the world's premier physicists witnessed "Trinity"—code name for the first detonation of a nuclear weapon, a plutonium implosion bomb near White Sands, New Mexico. After the initial detonation flash, Fermi began releasing wads of paper into the air and watched as they were propelled several meters by the first shock wave. From this simple experiment, lasting just a few seconds, he was able to estimate the yield of the world's first atomic explosion, which turned out to be within a factor 2x of the final correct number.

While our approximations here are for what I term "the biohealthcare industry" (BHI), the approach is similar to Fermi's: we try to break down problems to a linked series of reasonable assumptions or values we can estimate, then build them back up to identify the plausible range for an answer. Precision is typically not desired; obtaining answers within a factor of ten (also known as an order of magnitude) is all we need to make initial decisions. This may be an obvious step for those of you with an engineering, chemistry, or physics background—you already make an initial approximation before embarking on a more detailed analysis or effort to support a business plan or major project. However, this is less commonly used outside of the physical sciences, especially in the field of healthcare. Learning how to adopt Fermi's approach for your own initial estimates will quickly prove its value.

How is it done? Several authors describe the Fermi approach better than I can, but there is no one all-purpose algorithm. However, I will summarize some key points from two of my favorite writers on the subject: Lawrence Weinstein and Clifford Swartz. Weinstein in particular does a nice job of providing a high-level summary of the process.

Before we start, let's consider the reason for performing approximations. We estimate in order to inform potential decisions: how many doctors do we need to staff the clinic, how much will a national health plan cost, and so on. Before deciding to commit your time and resources to an effort, first determine the plausibility of a proposed enterprise. The Fermi method is ideal for this. You could say it is the ultimate "BS filter" to screen flawed ideas and projects before you waste much effort on them. You might be surprised by how infrequently people take this step. Given that many businesses fail in their first two years, teaching these simple plausibility estimates could be a real public service.

First step: Phrase the question of interest as specifically as possible. Second: Write down a plausible answer to a question, like "How many flower petals are on a stem?" Third: If you cannot immediately guess an initial answer within an order of magnitude, break the problem down into steps you can estimate individually, then combine them.

Sounds simple, right? It does take practice, interest, and perseverance to develop this skill, but it can be done.

A key sub-skill for Fermi estimation is the ability to estimate or approximate by using ranges. For example, it can be difficult to estimate the exact time it would take a nurse to check a patient into a room. However, with some experience in healthcare we can determine that it's unlikely to take less than one minute or more than an hour. Using this crude range (some might call it "worst-case bounds") we can then use the geometric mean to obtain a point estimate as follows:

$$Estimate = \sqrt{minimum \times maximum}$$

Why not use the arithmetic mean for the point estimate? This would work if our ranges were relatively close together. However, for many problems, our uncertainty range can be multiple orders of magnitude apart, such as completion time for complex activities. The result is that the arithmetic mean can be skewed by a broad range of values. In contrast, the geometric mean corresponds to the arithmetic mean of the logarithm of values and can easily accommodate wide ranges.

For example, if we are estimating a probability and our ranges are between 1% and 100%, the arithmetic mean would be approximately 50%—a mean that is fifty times the lower number and half the upper number. In contrast, the geometric mean would be 10%, or ten times the lower number and one tenth of the upper number. Thus, the geometric mean is the preferred method, because it gives us a number that is equal orders of magnitude apart.

Throughout this book, I typically compute the geometric mean using a calculator and rounding the numbers. That's a bit lazy, but it fits with the informal nature of estimation. You could also use a mathematic relation to work out an approximate geometric mean without a calculator or spreadsheet. The trick is to rewrite the numbers using scientific notation and take the arithmetic average of the coefficients and exponents. Using this method, our previous example would become:

$$\left(\frac{1+1}{2}\right) \times 10^{\left(\frac{0-2}{2}\right)} \approx 1 \times 10^{-1}(\%) \; or \; 10\%$$

This is not a perfect approximation, but it can be used in most situations if you don't want to use the calculator-and-computer approach.

Before moving to a discussion of units, I want to point out some pitfalls and types of problems that are not so easily handled by Fermi approximation (or at least not without care). The most difficult problems are those that are non-linear in nature: for example, queuing problems causing traffic jams, where a small number of additional cars on the road can suddenly bring things to a stop. I will explore this in more detail when we discuss both a linear approach to the time to inoculate a population and a queuing analytic approach to waiting in the doctor's office.

Another way the Fermi technique could skew results is by using uninformed point estimates for the inputs. It's crucial that you use plausible ranges or look up inputs online (or obtain from some reliable source) if possible. Simply "dumping" ill-informed numbers into a formula will produce useless information.

A third pitfall is using the wrong formula. It's good practice to include and cancel units (dimensional analysis) to ensure that your formulas at least provide values in the correct dimensions of interest. For example, if you're attempting to estimate the number of cars going through a lane over time, and the units of your formula result in something other than cars per unit time, then you have made a mistake.

When you are trying to solve a problem that falls into one of the above three categories, remember that practice will develop your ability to see when to be careful or when you have gone wrong.

CHAPTER 3

NOTATION AND NUMBERS

$$1{,}000 = 10^3$$
$$420 = 4.2 \times 10^2$$
$$.015 = 1.5 \times 10^{-2}$$
$$860 \text{ Billion} = 860 \times 10^9 = 8.6 \times 10^{11}$$
$$45 \text{ msec} = 4.5 \times 10^{-2} \text{ sec}$$

We are surrounded by numbers daily. We read in the news that the U.S. spent 17% of its GDP on healthcare last year or that the cost of treating a disease worldwide is approaching $7 billion dollars. The body contains a trillion cells; a drug costs hundreds of millions to get to the point of an FDA approval. The list goes on and on. What do all these numbers mean, and how should we think of them?

Fortunately, we can use an approach that summarizes numbers for easy calculation and to put them into context. This approach is known as scientific notation: we restate large numbers as a single leading number called the coefficient (sometimes with decimal places between 1 and 9.9), multiplied by a power of 10, called the exponent. For example, we can restate the approximate U.S. population in 2019, 300 million, as 3×10^8. Likewise, we could restate the approximate total U.S. debt obligation, about $70 trillion (including national debt, future healthcare, and social security obligations), as 7×10^{13}. For the purposes of this book, we will typically keep only one coefficient digit of precision, but in some cases we may add one decimal place. The more digits, the more implied precision of an estimated number. However, don't be fooled into thinking that precision implies accuracy!

I will use scientific notation for most problems in this book. This simplifies the math and allows us to perform basic algebra with ease. Even if you are familiar with manipulating large numbers, remember the following rules:

- Addition or subtraction: Add/subtract the coefficients adjusted to the same exponent. E.g.,
  $(1 \times 10^3) + (4 \times 10^1) = (1 \times 10^3) + (.04 \times 10^3) = 1.04 \times 10^3 \approx 1 \times 10^3$

- Multiplication: Multiply coefficients and add exponents. E.g., $(2 \times 10^3) \times (3 \times 10^1) = (2 \times 3) \times 10^{(3+1)} = 6 \times 10^4$

- Division: Divide coefficients and subtract exponents. E.g., $(2 \times 10^3) / (3 \times 10^1) = 2/3 \times 10^{(3-1)} = 0.67 \times 10^2 \approx 7 \times 10^1$

In most cases, I will limit the coefficient to a single digit, since the goal is to achieve accuracies on the order of 10x (one order of magnitude). Occasionally, I will keep one extra digit of precision to allow us to complete the calculation, especially with small numbers of zero magnitude, such as 100 power. This is largely due to habit on my part; you do not have to follow this approach. You can go through all problems and use only powers of 10 and still achieve quite satisfactory results.

If you want additional instruction on working with scientific notation, I recommend looking at either of Weinstein's Guesstimation books.[3] He provides an excellent review and discussion on dealing with large numbers, along with several worked examples.

---

[3] Lawrence Weinstein and John A. Adam, *Guesstimation: Solving the World's Problems on the Back of a Cocktail Napkin* (Princeton: Princeton University Press, 2008); Lawrence Weinstein, *Guesstimation 2.0: Solving Today's Problems on the Back of a Napkin* (Princeton: Princeton University Press, 2012).

CHAPTER 4

THE METRIC SYSTEM, UNIT CANCELLATION, AND CONSTANTS

$1 \text{ inch} = ? \text{ feet} = ? \text{ yards} = ? \text{ miles}$

$1 \text{ mm} = .01 \text{ cm} = 1 \times 10^{-3} \text{ m} = 1 \times 10^{-6} \text{ km}$

OMG, how do I convert units?

The metric system (meters-kilogram-seconds, or MKS) is ideal for calculations of all kinds, including estimation problems. You will see throughout this book that I may take U.S. or U.K. standard units and convert them to MKS units wherever possible. In some cases, I convert these back to standard U.S. units for the reader. Consequently, I have adopted the habit of cancelling units to ensure that my estimates are at least giving me the appropriate dimensions. I highly encourage you to use the same approach and to use symbols for your estimates until the last step or so; you can substitute in your individual estimates to complete the calculation. Let me illustrate the unit-cancellation technique with an example.

Assume you are trying to estimate the area of your yard to order sod replacement. Using your feet, you estimate the length of the yard to be about 82 steps long and 48 steps wide. Assuming that your steps are approximately equivalent to a one-foot measure, you can calculate the area of the lot as:

$$Lot\ area = 82\ ft \times 48\ ft \approx 3{,}900\ ft^2$$

Unfortunately, the company providing sod sells it in pallets based on square meter area. How do you determine your lot area in meters? You can easily use a Google unit-conversion application, but you likely know that one meter equals roughly three feet. We can convert and cancel units to arrive at the lot area of:

$$Lot\ area\ (m^2) = 3{,}900\ \cancel{ft^2} \times \frac{1\ m^2}{9\ \cancel{ft^2}} \approx 400\ m^2$$

Note that I used a fairly crude conversion here, based on my experience and keeping with simple order-of-magnitude estimates. I rounded the result to 400 square meters for illustration purposes. However, if I only have a limited budget for landscaping and the cost of sod could bust my budget if I'm not careful, I would pursue a more precise estimate by carefully measuring the lot dimensions and using conversion factors with more digits of precision. This illustrates a key point about approximations: Your initial estimates result in a high-level picture of the answer to a problem. Depending on the results of your initial estimation, you may need to do more precise calculations and measurements to further reduce the uncertainty.

Finally, you may notice that I sometimes refer to known constants: acceleration of gravity, conversion of pounds to kilograms, prevalence of epilepsy in the U.S., and so on. The ubiquity of online search tools or math and physics tables makes memorizing them unnecessary. I recommend that you look up constants and conversions as needed. This will save time and allow you to focus on the more important and fun aspects of estimation.

CHAPTER 5

FROM RANGES TO DISTRIBUTIONS

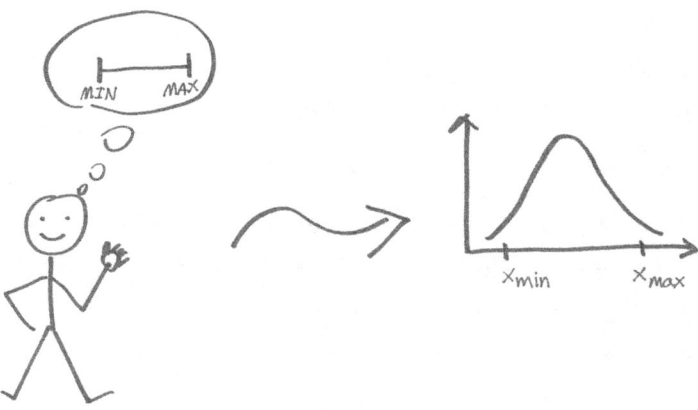

Note: You can skip this chapter and the next if you're not a math geek and just want to move on. You'll still be able to understand and follow the examples in the book without it.

For some problems, it's necessary to produce estimates with some characterization of the likely ambiguity, beyond assuming an order-of-magnitude uncertainty. For example, in some Bayesian calibration tests, experts are asked to produce subjective range estimates like the following: "What is the height of the tallest building on earth? Please provide your

90% confidence interval for the range of likely values."

Here you are asked to use your background knowledge about the world to provide a bounded estimate that you are 90% confident contains the correct value. In other words, we expect to be wrong 10% of the time when our estimates are evaluated over a number of different questions. Note that this concept is subtly different from the frequentist statistical concept of a confidence interval ("CI"), where the 90% interval is a range that contains the true mean—calculated from multiple random experiments—90% of the time.

So how would we go about calculating such an interval? Here's a crude but useful approach that merges together concepts from estimation, statistics, and project management. In the 1950s, the U.S. Navy commissioned a report to develop a method for estimating activity and completion times for projects. This process was termed the "Program Evaluation and Review

> **TWO KINDS OF PROBABILITY**
>
> For those of you unfamiliar with the field of probability and statistics, you might be surprised to hear that there are two general interpretations of probability. Let me briefly explain the differences. The first approach views the probability of an event or outcome as the relative frequency of outcomes from a random variable over many random experiments.
>
> For example, the probability of "heads" for a fair coin would be the long-run proportion or frequency of "heads" observed with a very large number of coin-flipping experiments. This is often called the "frequentist interpretation" or "empirical probability." Another interpretation views probability as a personal "state of knowledge" or degree of belief about the likely outcome(s) of an event or quantity. This is known as the subjective probability interpretation.
>
> For this book, you will note that I freely use both frequentist and subjectivist approaches whenever convenient, such as when I use Bayes' theorem to assess the probability that a diagnosis is correct. If you're interested, you can read a long history of mathematical arguments and discussions from proponents of both forms. I freely acknowledge that I am an eclectic, equal-opportunity user of both frequentist and subjectivist methods. For a more detailed discussion of this topic, see Doug Hubbard's book on measuring intangibles in business.
>
> Douglas W. Hubbard, *How to Measure Anything: Finding the Value of Intangibles in Business*, 3rd ed. (Hoboken, NJ: John Wiley & Sons, Inc., 2014).

Technique," or PERT. The PERT approach requires subject-matter experts to provide three time estimates for tasks, including minimum and maximum. Presumably, all followed a mathematical beta function distribution shape. Although this technique was originally developed in 1958, it is still one of the cornerstone algorithms for time estimation used in project management today.

For mathematical ease, I will use an even simpler approach. I'll assume we only have the minimum and maximum values (our typical Fermi ranges), along with the assumption that our distribution is typically skewed but becomes normal when transformed to logarithms. Thus, I assume that most of our estimates will be log-normal (the log values follow a Gaussian distribution, or a bell-shaped curve) in character. The results will actually end up being similar to those of the PERT assumptions of a beta distribution but will be somewhat easier to calculate. This is essentially what we are doing when we calculate the geometric mean, as I explain below.

Let me begin with an example to illustrate the concept and approach. First, let's assume we are estimating parameters α, β, and δ, which define the extreme limits and midpoint of a normal distribution in natural log space, as illustrated below:

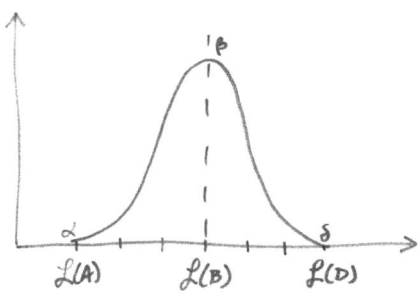

Figure 1. A log-normal distribution in log space

The three parameters are all logarithmic transformations of the estimates from linear space:

$$\alpha = Ln(A)$$
$$\beta = Ln(B)$$
$$\delta = Ln(C)$$

These "log-normal" quantities are plotted in linear space as a right-skewed probability distribution as shown below:

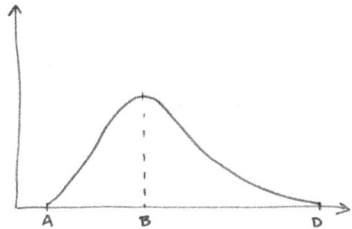

Figure 2. The same distribution in linear space

Using the distribution in Figure 1.0, we can estimate the mean value (β) as the midpoint between the extremes, or:

Equation 1.0: $\beta = \frac{\alpha + \delta}{2}$

While not immediately apparent, this is the same result we could obtain by calculating the geometric mean using the linear (not log-transformed) values of A and B as the extremes of our range with Fermi approximation. A point to remember is that for log-normal data, the geometric mean is equivalent to the median. Here we see that the arithmetic mean of the log extremes is the same as the geometric mean of the linear extremes. I will generally use the term "mean" but am assuming that this is really the *geometric* mean derived using the square root rule introduced in Chapter 2, or by taking the average of the logs as shown above.

Now we want to estimate a quantity known as the standard deviation. Here I will use the symbol σ, which is a common notation for the population deviation or measure of variability about the mean that is appropriate here, assuming the ranges reflect the plausible population ranges. Finally, I make the assumption that we have chosen our extreme values (A and C) such that the corresponding logarithmic transformations (α and β) cover at least 99% of the possible distribution. Assuming that what we are estimating is well characterized by a unimodal (one central tendency or peak) normal distribution in log space, we can say that the difference in the extremes are approximately six standard deviations, or 6σ. This is illustrated below in Figure 3.0, using a standard normal distribution with z values. With this observation and assumptions, we can solve for our estimated standard deviation as follows:

$$\text{Equation 2.0:} \quad \sigma = \frac{(\delta - \alpha)}{6}$$

Finally, once we know our mean value (β), we can then provide a rough estimate for our 90% CL (from the 5% confidence limit to the 95% confidence limit) as follows:

$$\text{Equation 3.0:} \quad 90\% \; CL \sim \beta \pm 2\sigma$$

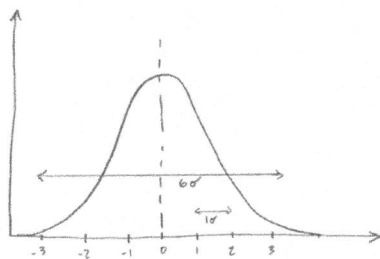

Figure 3. 6σ illustration

Another way to think of this is that four standard deviations centered about the mean cover ≈ 90–95% of the distribution. This gives us our approximate range. I have chosen two as our multiplier (while the actual 95% z-value = 1.96), which in the spirit of approximation should be close enough for our purposes. Finally, we must convert the results from Equation 3.0 back into our conventional (linear) units, using the identity relationship between logarithms and exponentials:

Equation 4.0: $B = e^{\beta}$

Equation 5.0: *Lower 5% CL* $\approx e^{\beta - 2\sigma}$

Equation 6.0: *Upper 95% CL* $\approx e^{\beta + 2\sigma}$

Note that I have performed a number of mathematical steps to reach this point. Why couldn't I just forget the logarithmic transformations and compute the geometric mean and standard deviation using our conventional/linear units? The reason is that in many cases, our plausible estimates are greater than an order of magnitude. This can lead to skewed distributions like the one in Figure 2.0, and simply estimating the confidence interval and computing confidence limits with Equation 2.0 can lead to negative lower-confidence limits which are nonsensical in most cases. (Imagine a negative time to complete a task—it makes no sense.) Using logarithms avoids an illogical result, although with the cost of a little extra mathematical work.

To illustrate this approach, I will estimate the height of the tallest building on earth. Based on my experience, I know that the Empire State Building in New York City is at least about 1,000 feet in height (100 floors, with about 10 feet per floor). I also know that many planes require oxygen at 10,000 feet, and I've never seen a building that tall with a base at sea level or where the occupants require oxygen. Thus, for this example I will use my prior knowledge to estimate a height between 1,000 and 10,000 feet using my previous approach.

Here, we let A = 1,000 ft and C = 10,000 ft. Converting to natural logarithms gives us:

$$\alpha = Ln(A) \approx 7 \text{ and } \delta = Ln(C) \approx 9$$

$$\beta = (7 + 9)/2 \approx 8$$

$$\sigma = (9 - 7)/6 \approx 0.4$$

Using these values, we can first solve for our approximate 90% confidence limits (CL) in log space:

Lower 5% CL ≈ 8 - (2 × 0.8) ≈ 7.2

Upper 95% CL ≈ 8 + (2 × 0.8) ≈ 8.8

These values are then finally converted back to our conventional, linear units:

Lower 5% CL ≈ $e^{7.2}$ ≈ 1,300 ft

Upper 95% CL ≈ $e^{8.8}$ ≈ 6,600 ft

Geometric mean ≈ e^{8} ≈ 2,900 ft

Note that I have rounded calculations to one or two significant figures, so your calculations may differ slightly depending on how many decimal places you keep. A quick glance at Wikipedia tells me that the height of the Empire State Building is between 1,250 and 1,454 feet, depending on whether you only look at roof height or count the tower and antenna. Based on this information, my lower 90% limit contains the upper value, but almost misses the lower building height value by 50 feet. However, we are looking for the range containing the tallest building in the world, not the height of the Empire State Building. The Wikipedia list of building heights on October 30, 2019, shows that the current honor for world's tallest building goes to the Burj Khalifa in Dubai, at 828 m (2,171 ft). This figure falls within our estimated 90% confidence interval.

Finally, the following graphs illustrate both the linear and lognormal probability distributions and the skewing of the confidence intervals for this example.

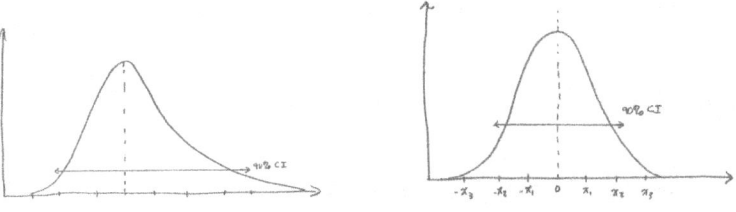

Figure 4. Skewed distributions and confidence intervals

In summary, I have gone a long way to illustrate what I admit is a crude approach to estimating uncertainty. However, doing this is sometimes useful for simple situations, and it's important to be clearly aware of the assumptions and cautions if you want to use this approach. My results will often be reasonably close to those obtained using the PERT method, and my estimation of the standard deviation uses similar reasoning.* This method takes a few more calculations than some of the other examples in this book, but it allows you to provide a plausible boundary on the associated uncertainty.

---

* A caveat: Converting the log-normal mean and standard deviation to the normal linear space equivalents requires additional adjustments that I have not performed here. The differences between my simple conversion and the exact methods do not usually result in large discrepancies, but be aware that I am taking shortcuts with the math in the spirit of approximation.

CHAPTER 6

UNCERTAINTY PROPAGATION

Note: If you want to skip this chapter, you will still be able to understand and follow the examples in the rest of the book.

In the previous chapter, I introduced an approach for estimating uncertainty for a range derived for use with Fermi approximation. How would one estimate the uncertainty for multiple quantities when they are operated on (multiplied, for example)? Below I will present a simple analytical method to handle these situations, drawing on the work of my former physics professor Dr. Clifford Swartz from his book *Used Math*.[4]

---

[4] Clifford E. Swartz, *Used Math for the First Two Years of College Science*, 2nd ed. (College Park, MD: American Association of Physics Teachers, 1993).

Using the Fermi method, assume we have estimated two quantities of interest as $x \pm \sigma_x$ and $y \pm \sigma_y$, where x and y are the geometric means and σ_x and σ_y are the associated standard deviations, respectively. Given these values, we can define the resulting uncertainty as the aggregate standard deviation from three operations:

1. For addition and subtraction: $x \pm y$ the associated uncertainty $\sigma_{x \pm y}$ is given by $\sqrt{\sigma_x^2 + \sigma_y^2}$
2. For multiplication: $x \times y$ the associated uncertainty σ_{x*y} is given by $\sqrt{\left(\frac{\sigma_x}{x}\right)^2 + \left(\frac{\sigma_y}{y}\right)^2}$
3. Finally, for division: $x \div y$ the associated uncertainty $\sigma_{x/y}$ is given by $\sqrt{\left(\frac{\sigma_x}{x}\right)^2 + \left(\frac{\sigma_y}{y}\right)^2}$

A word of caution: As the complexity of the problem increases, the required calculation steps quickly become tedious. Thus, I recommend this approach for simple cases where one is dealing with no more than two or three quantities, otherwise the manual bookkeeping effort and chance for making arithmetic errors escalates quickly. For complicated cases with multiple calculation steps where you need assessments of the uncertainty, I would recommend using a simulation method such as Monte Carlo to create estimates. Monte Carlo simulation is a computational method that randomly varies the inputs to a mathematical model and samples the results to obtain solutions. I will discuss more about this in Chapter 41, "Bayes Meets Shakespeare's Monkeys: Monte Carlo Bayesian Analysis."

To illustrate the utility of propagating uncertainty, suppose you are evaluating bids from external vendors to write safety narratives for clinical trials. You receive estimates from your internal team, suggesting that it can take between one and six hours to write and quality-check a narrative document, and typical vendor costs range between $10 and

$20 per hour. Using this information, what would you estimate as the 90% CL for total costs to produce 5,000 narratives?

We begin by estimating the geometric mean and the PERT standard deviation for our time and cost quantities, using the approaches discussed so far. For illustration purposes, I will keep quantities to one decimal place of precision:

$$T_{writing} = \sqrt{1 \times 6} \; hrs/doc \approx 2.5 \; hrs/doc; \text{ and}$$

$$\sigma_t = \frac{1-6}{6} \approx 0.9 \; hrs/doc$$

$$C_{writing} = \sqrt{10 \times 20} \; \$/hr \approx 14.1 \; \$/hr; \text{ and}$$

$$\sigma_c = \frac{10-20}{6} \approx 1.7 \; \$/hr$$

The total estimated cost per document is the multiplication of the mean writing time per document and mean writing cost per hour. We then estimate the pooled relative standard deviation using our formula:

$$\sigma_{cost} = \sqrt{\left(\frac{\sigma_x}{x}\right)^2 + \left(\frac{\sigma_y}{y}\right)^2} \; ; \text{ which } = \sqrt{\left(\frac{0.9}{2.5}\right)^2 + \left(\frac{1.7}{14.1}\right)^2} = 0.36 \text{ or } 36\%$$

Now we use these results to determine the approximate upper and lower 90% limits for the cost per document:

$$Mean \; cost \approx 2.5 \; \frac{hr}{doc} \times 14.1 \frac{\$}{hr} \approx \$35 \; per \; document, \sigma_{cost} = 0.36 \times \$35 \approx \$13$$

$$Lower \; 90\% \; CL \approx \$35 - (1.65 \times \sigma_{cost}) \approx \$14 \; per \; document$$

$$Upper \; 90\% \; CL \approx \$35 + (1.65 \times \sigma_{cost}) \approx \$56 \; per \; document$$

Our estimated cost per document with the associated 90% CL then becomes approximately $35 ($14, $56). Since there is no uncertainty associated with the 5,000 narratives (because we specified this), we would expect that vendors should provide cost estimates that are within

10–20% of our total estimate of $175K ($70K, $276K). If the vendor proposal is significantly different from our estimate, you can quickly ask questions to understand why.

CHAPTER 7

HOW MANY JELLYBEANS IN A JAR?

When I was a boy, I often saw guessing contests at state fairs and school fall festivals. The classic one I remember was guessing how many jellybeans were in a large jar. This is a good real-life example of where estimation can be useful, and it makes a very good introduction to Fermi approximation.

The general approach is to estimate the number of jellybeans by calculating the ratio of the jar volume by the average jellybean volume. We can assume that both the holding jar and jellybean shape can be modeled as cylinders. Remember, we're looking for approximate answers, not exact ones. Jars are largely cylinders with some decorative aspects such as jar necks, but we'll ignore those in our calculations. Jellybeans are really more ellipsoidal in shape, but to keep the math easy we will use a cylinder. We could even use spheres or cubes in our calculations, but let's stay with cylinders. I've drawn out the general cylinder geometry and formula for volume below:

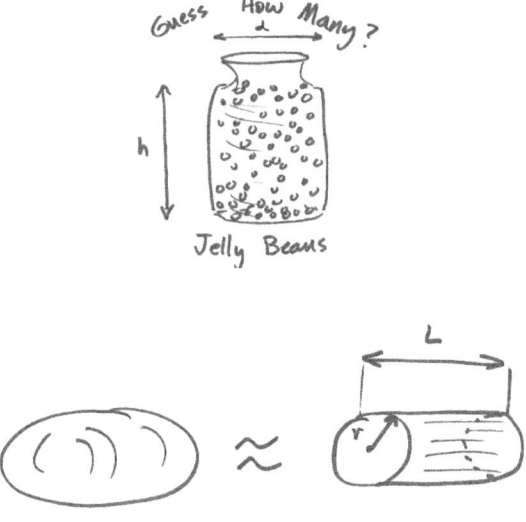

Figure 5. Cylinder volume (units³) = $\pi r^2 h$ (or l)

Assuming the jar is full, and we ignore the neck and mouth of the jar, we need to obtain estimates for a few parameters:

1. What is the height of the jar and average jellybean? Call these parameters H_{jar} and H_{jb}.

2. What is the radius of the jar and average jellybean? Call these parameters R_{jar} and R_{jb}, where R is half the diameter (D).

We could now proceed with estimating these values, but I want to introduce one more term. Picture a box of marbles or small rocks. When things are packed together, there is often some space between them due to the geometry. These empty spaces cannot be eliminated, so for more advanced calculations you need to use a parameter known as a "packing factor" (ρ) or "packing efficiency." This parameter has a value between zero and one, with one indicating that objects can be packed together with little or no empty space between them and zero indicating that the objects have maximal empty space between them. Thus, we will want to estimate and use this parameter, too.

First, let's write down our general formula for the number of jellybeans:

$$N_{jb} = \frac{\rho V_{jar}}{V_{jb}}$$

Next, assume the dimensions for the holding jar. Based on my (admittedly flawed) recollections from childhood, I remember very large jars about two feet in height and two feet in diameter (a radius of one foot). To make the math easier, let's convert these to metric units (1 inch ≈ 3 cm) that are used for scientific calculations:

$$H_{jar} = 2 \text{ ft} \times \frac{12 \text{ in}}{\text{ft}} \times \frac{3 \text{ cm}}{\text{in}} \approx 72 \text{ cm}$$

$$R_{jar} = 1 \text{ ft} \times \frac{12 \text{ in}}{\text{ft}} \times \frac{3 \text{ cm}}{\text{in}} \approx 36 \text{ cm}$$

I'm going to assume that a jellybean is about one inch in length and about half that in diameter (radius of half an inch). Convert these dimensions to metric units:

$$H_{jb} = 1 \text{ in} \times \frac{3 \text{ cm}}{\text{in}} \approx 3 \text{ cm}$$

$$R_{jb} = \frac{1}{2} \text{ in} \times \frac{3 \text{ cm}}{\text{in}} \approx 1.5 \text{ cm}$$

Estimate a packing factor:

$$\rho = \sqrt{.5 \times 1.0} \approx .7$$

Estimate the respective volume of the jar and jellybean:

Jar volume $(cm^3) = \pi \times 36 \text{ cm}^2 \times 72 \text{ cm} \approx 300 \times 10^3 \text{ cm}^3$

Jellybean volume (cm³) $= \pi \times 1.5\ cm^2 \times 3\ cm \approx 21\ cm^3$

Use the packing factor to complete the estimate for the number of jellybeans:

$$N_{jb} = \frac{0.7 \times 300 \times 10^3\ \cancel{cm^3}}{21\ \cancel{cm^3}} \approx 10{,}000\ jellybeans$$

This is much more than I would probably have guessed at the time! If I had used this technique, I would have at least produced a plausible estimate within a factor of 10x. Note that I could have also created a plausible range for my uncertainty using the techniques in the previous chapters, although the math would have been a little tedious.

CHAPTER 8

AROUND THE WORLD IN 80 DAYS?

In the nineteenth century, Phileas Fogg, the protagonist of Jules Verne's famous novel,[5] took up the challenge to circumnavigate the earth in 80 days and return to London to claim his victory. How long would such a trip have taken in the late 1800s? This turns out to be another good Fermi estimation task. As with the jellybeans problem, we need to make some plausible assumptions to proceed.

Let's start by estimating the circumference of the earth. Rather than looking this up on the Web, let's use the general knowledge that the earth is divided into 24 time zones, and each zone is about 1,000 miles'

---

[5] Jules Verne, *Around the World in Eighty Days* (Paris: Pierre-Jules Hetzel, 1873).

distance at the equator. We can thus assume the earth is about 24,000 miles in diameter.

The next step is estimating Mr. Fogg's travel speed. In the novel, Mr. Fogg and company walked, rode elephants, took a train, and sailed on ships throughout their journey. Consulting the novel, we see that the itinerary he planned consisted of railway and steamer passage. At one point in the journey, the company had to unexpectedly travel by elephant due to problems with rail travel through rural India.

Let's begin our calculations by estimating the rate of travel for each method. In the early nineteenth century, many people believed that human beings could not survive speeds greater than 20 miles per hour. This myth was dispelled by the time of Verne's novel, so railway speeds achieving 60 miles per hour or more were not uncommon. Let's estimate rail travel at a lower speed of 20 to the upper limit of 60:

$$Rail_{vel} = \sqrt{20 \times 60} \approx 35 \text{ miles per hr}$$

Travel by steamer was much slower than rail travel. I estimate the steamer's speed as faster than walking, at the lower end of the scale, but no faster than human sprinting, at 25 miles per hour:

$$Steamer_{vel} = \sqrt{5 \times 25} \approx 11 \text{ miles per hr}$$

Travel by elephant is a little less certain. I estimate its speed to range between that of a walking human (three miles per hour) to less than the steamer maximum:

$$Elephant_{vel} = \sqrt{3 \times 15} \approx 7 \text{ miles per hr}$$

Next, we consider the daily travel time. While on the steamer, Fogg and company traveled continuously, since boats don't stop until they reach land. The same is true for train travel. Travel by elephant, however, would involve starts and stops and only accounts for a small portion of the story. In addition, the story characters had to occasionally stop due to delays or to switch transport, obtain passes, and so on. Thus, we should estimate a daily travel time that is less than 24 hours per day. Let's assume that the shortest travel days would be by elephant and could last up to 12 hours, and the longest travel days would be by train and steamer at 24 hours. Now we must estimate the relative duration of each segment of the trip, but we'll need to take into account the amount of time spent on each mode of travel. This calculation would affect both our velocity calculations and travel time per day.

Let's do this first by using only the rail and steamer velocities. We know that approximately 75% of the earth is covered by water, so I will assume that this leaves only 25% of the overall distance spent on land travel. If the pachyderm detour through India accounts for only a small part of the trip (say ~ 5%), then we can create an estimated aggregate velocity and travel duration as:

$$Avg_{vel} = .75 \times \frac{11 \ mi}{hr} + 0.25 \times \frac{35 \ mi}{hr} \approx 17 \frac{mi}{hr}$$

Our travel time in days thus becomes:

$$Time_{days} = \frac{24{,}000 \ \cancel{mi}}{trip} \times \frac{1 \ \cancel{hr}}{17 \ \cancel{mi}} \times \frac{1 \ day}{17 \ \cancel{hr}} \approx 83 \frac{days}{trip}$$

Ok, this calculation is very close to the book's timeline. What would happen if we also account for the slower travel by elephant? Assuming that the pachyderm land travel through India accounted for 1% of the trip, we could simply re-calculate the average velocity as:

$$Avg_{vel} = \left(0.75 \times \frac{11 \ mi}{hr}\right) + \left(0.20 \times \frac{35 \ mi}{hr}\right) + \left(0.05 \times \frac{7 \ mi}{hr}\right)$$
$$\approx 16 \frac{mi}{hr}$$

This makes the overall trip time:

$$Time_{days} = \frac{24{,}000 \ \cancel{mi}}{trip} \times \frac{1 \ \cancel{hr}}{16 \ \cancel{mi}} \times \frac{1 \ day}{17 \ \cancel{hr}} \approx 88 \frac{days}{trip}$$

This is certainly longer than our first approximation, but still very close to the book's projected time (< 20% error). We can also simply estimate the average velocity using the lowest elephant speed and highest train speed as follows:

$$Avg_{vel} = \sqrt{3 \times 60} \approx 13 \text{ miles per hr}$$

This gives us a final estimated duration of:

$$Time_{days} = \frac{24{,}000 \cancel{mi}}{trip} \times \frac{1 \cancel{hr}}{13 \cancel{mi}} \times \frac{1 \, day}{17 \cancel{hr}} \approx 108 \frac{days}{trip}$$

This longer trip duration is only 20% more than Fogg's actual trip and is well within the factor of a 10x difference that we attribute to Fermi approximation. Not bad! In summary, I would guess that a realistic time to accomplish Fogg's trip would be between about 80 and 110 days, and I would have been among the club members betting against his success.

CHAPTER 9

HOW MANY DRIVERS ARE TEXTING RIGHT NOW?

To estimate the answer, we must break this problem into two components: the number of drivers right now and the number of drivers texting right now. We can solve this and many other similar problems by using the following insight: the proportion of people performing an activity is equivalent to the proportion of time spent doing the activity.

We'll use this understanding to first estimate the number of drivers right now. As an average U.S. nonprofessional driver, I rarely drive anywhere for less than thirty minutes. Even my quickest errands take that time or longer. Some days I have spent several hours driving, depending on the commute and associated appointments and errands. I estimate the upper end of my driving time as three hours. We can use this estimation to calculate the geometric mean time per day of driving:

$$T_{drive} = \sqrt{.5 \; hrs \times 3 \; hrs} \approx 1.2 \; hrs \; per \; day$$

What proportion of the day is this?

$$P_{time} = \frac{1.2 \; hrs}{24 \; hrs} \approx 5\%$$

Next, we need to estimate how many people are active drivers. Using the same approach, I can assume that at least 50% of the population is driving. This gives me an estimate of:

$$P_{drive} = \sqrt{50\% \times 100\%} \approx 70\%$$

Assuming the U.S. population is 300 million, we can estimate the average number of drivers as:

$$N_{drivers} = 5\% \times 70\% \times (300 \times 10^6) \approx 10 \times 10^6 \; drivers$$

Using the same method, we can now estimate the number of drivers texting. I probably spend more than five minutes a day texting, but my maximum amount may be unrepresentative. However, I estimate that few individuals can actively text for more than three hours per day. This gives us an estimate of:

$$T_{text} = \sqrt{5 \; min \times 180 \; min} \approx 30 \; mins \; per \; day$$

This is half an hour, which is not a big part of the day:

$$P_{text} = \frac{0.5 \; hrs}{24 \; hrs} \approx 2\%$$

Assuming the U.S. population is 300 million, we can estimate the average number of U.S. drivers who are texting as:

$$N_{drivers} = 2\% \times (10 \times 10^6 \; drivers) \approx 200{,}000 \; drivers$$

Not a huge number, but enough to cause slowdowns and accidents. Don't text and drive!

CHAPTER 10

SHAKESPEARE'S MONKEYS

While you may never have been formally introduced to the "Infinite Monkey Theorem," you will recognize the key point of this idea: randomness can generate patterns of significance. Historically, this idea was popularized with the rise of statistical mechanics. The great mathematician Emile Borel and others theorized that large numbers of typing monkeys, given a random generation of letters and enough time, could reproduce many of the works of great literature. (This hypothesis applies to machines as well as to biological monkeys.) In addition to Weinstein and Adam (cited in Chapter 3), Aaron Santos has suggested several nice variations of and solutions to this problem.[6]

Let's look at Borel's idea in detail. Assume we have one million hypothetical digital "monkeys" that can bash away on typewriters with the 26 keys of uppercase letters of English, eight hours a day, with no rest. How we keep our monkeys interested and focused on this task is another question. However, we'll assume they are happy to be engaged in this exercise and can strike a key every second. How long will it take them to reproduce the following phrase? "To be, or not to be…"[7]

We have 13 letters of English with five spaces (which we will ignore). How would we calculate the probability of getting all 13 letters exactly correct? Knowing there are 26 ways to obtain each letter in the English/Latin alphabet by random chance, we can write the expression for getting the first letter correct as:

$$Probability = \frac{1}{26}$$

In order to generate all 13 letter matches, assuming that each is produced independently, we would simply multiply each of the individual letter probabilities by 13, giving us:

$$Probability\ (13\text{-letter Shakespearean phrase}) = \left(\frac{1}{26}\right)^{13}$$

This turns out to be a very small number indeed: 4×10^{-19}. The reciprocal of this gives us the estimated number of keys the monkeys must hit to generate these 13 letters: 2.5×10^{18}. To get an idea of how big this number is, calculate that if we have one million monkeys typing one letter per second, it would take:

$$Typing_{rate} = 10^6\ \cancel{monkeys} \times \frac{1\ letter}{sec - \cancel{monkey}} \approx 10^6\ letters/sec$$

---

[6] Aaron Santos, *How Many Licks? Or, How to Estimate Damn Near Anything* (Philadelphia: Running Press Book Publishers, 2009).
[7] William Shakespeare, *Hamlet*, Act 3, Scene 1.

$$Time = 2.5 \times 10^{18} \; \cancel{letters} \times \frac{1 \; \cancel{sec}}{10^6 \; \cancel{letters}} \times \frac{1 \; year}{31.6 \times 10^6 \; \cancel{sec}}$$

$$\approx 79{,}000 \; years!$$

What if we used one million computers instead of monkeys, with each generating 10^6 letters per second? Then our calculations speed up quickly and become:

$$Time = 2.5 \times 10^{18} \; \cancel{letters} \times \frac{1 \; sec}{10^{12} \; \cancel{letters}} \times \frac{1 \; hr}{3{,}600 \; \cancel{sec}}$$

$$\approx 700 \; hours!$$

The question becomes more relevant if we substitute "passwords" for Shakespeare's lines. Assume we are no longer limited to 26 upper or lowercase letters of English but have 26 upper + 26 lower + 10 digits + 15 characters, or 77 possible characters for a possible password. Then the probability of guessing all letters correctly for a 13-digit password becomes:

$$\textit{Probability (password letter)} = \left(\frac{1}{77}\right)^{13} = 3 \times 10^{-25}$$

Now, even with one million computers, each processing one million letters (or characters) per second, the time it would take to guess the password becomes:

$$Time = 3.3 \times 10^{24} \; \cancel{characters} \times \frac{1 \; \cancel{sec}}{10^{12} \; \cancel{characters}}$$

$$\times \frac{1 \; year}{31.6 \times 10^6 \; \cancel{sec}} \approx 100{,}000 \; years$$

If the password is slightly increased in length to just 15 characters, the time then becomes:

$$Time = 2 \times 10^{28} \; \cancel{characters} \times \frac{1 \; sec}{10^{12} \; \cancel{characters}}$$

$$\times \frac{1 \; year}{31.6 \times 10^6 \; \cancel{sec}} \approx 630 \; million \; years!$$

Hmmm. Who knew that a famous thought experiment from the past could have so much relevance today? Maybe it's time to update our passwords! Also, as we can see from the math, the numbers involved in generating the actual works of Shakespeare using random "monkeys" becomes truly larger than astronomical and would take much more than all the protons in the visible universe over multiples of its lifespan. It's true that given enough time and "monkeys," generation of these works is certain to occur. However, unless we are using a quantum computer to generate them, we will not be around to read them.

CHAPTER 11

HOW MANY BREATHS IN A LIFETIME?

I t's interesting to realize that much of human behavior is automatic and only partly under conscious control. Breathing, for example, is a critical function, but many people never really pay attention to it. Focus on your breath for a moment. How many breaths are in an average life? Millions? Gazillions? Well, let's estimate it! To answer this question, we need to either estimate or look up several pieces of information. First, what is the average breathing rate? Call this breathing frequency B_f. Second, what is the average human life span? Call this life span L_d.

You can time your own breath or look up the range from a standard medical reference. From previous experience, I know that the breathing rate can range from about three breaths per minute (experienced meditators) to about twenty breaths per minute (moderate physical exertion). Sustained breathing rates beyond this are less common and can be indicators of dangerous clinical conditions. For our purpose, this should be a reasonable range. Using the geometric mean formula, we can estimate the average as:

$$Breath_{freq} = \sqrt{3 \times 20} \approx 8 \; breaths \; per \; min$$

Now let us estimate the average human life. Assuming normal modern lifespans, I will use 60 for the lower boundary and 100 for the upper boundary. This results in:

$$Life_{duration} = \sqrt{60 \times 100} \approx 77 \; years \; per \; life$$

Since we know the number of days per year, hours per day, and minutes per hour, we can estimate the average number of breaths over a life span:

$$N_{breath} = life_{duration} \times days_{year} \times hr_{day} \times mins_{hr} \times breath_{freq}$$

Using our estimated values and canceling units, we have our results:

$$N_{breath} = \frac{77 \; \cancel{yr}}{life} \times \frac{365 \; \cancel{days}}{\cancel{yr}} \times \frac{24 \; \cancel{hrs}}{\cancel{day}} \times \frac{60 \; \cancel{mins}}{\cancel{hr}} \times \frac{8 \; breaths}{\cancel{min}}$$

$$\approx 3 \times 10^8 \frac{breaths}{life}$$

300 million breaths! Hard to believe until you do the math.

CHAPTER 12

HOW MANY HEARTBEATS IN A LIFETIME?

Breathing is not the only semi-automatic function we depend on. The heart pumps oxygen from our breath throughout the body and completes the circuit by delivering CO_2 to our lungs to be released. Using the same type of approach we used to calculate breaths, let's estimate the number of heartbeats in a life. This time we only need to estimate the average heart rate, since we have just estimated the average human life.

Let's call the average heart rate $heart_{rate}$.

You can time your heart or look up the range from a standard medical reference. Standard rates for adults (the rate for young children can be much faster) range from about 50 beats/minute (endurance athletes) to about 90 beats/minute. Rates much beyond this are considered tachycardia and are typically seen with exertion or clinical disease. Using the geometric mean formula, we can estimate the average:

$$Heart_{freq} = \sqrt{50 \times 90} \approx 67 \; beats \; per \; min$$

Now we can estimate the average number of heartbeats over a life span, using our previous life span estimate and similar time conversions:

$$N_{beats} = life_{duration} \times days_{year} \times hr_{day} \times mins_{hr} \times heart_{rate}$$

Using our estimated values and canceling units, we have our results:

$$N_{beats} = \frac{77 \; \cancel{yr}}{life} \times \frac{365 \; \cancel{days}}{\cancel{yr}} \times \frac{24 \; \cancel{hrs}}{\cancel{day}} \times \frac{60 \; \cancel{mins}}{\cancel{hr}} \times \frac{67 \; beats}{\cancel{min}}$$

$$\approx 3 \times 10^9 \frac{beats}{life}$$

This is about an order of magnitude more than our breaths, or approximately three billion beats. It's amazing to realize that our hearts can go for this long. A feat of nature indeed!

CHAPTER 13

HOW MANY HOURS OF SLEEP IN A LIFETIME?

OK, we've estimated the number of breaths and heartbeats in a lifetime. We can't forget sleeping! It's one of life's blessings and we do spend a lot of time doing it. (At least *I* do…)

Let's use the same approach to examine how long we hibernate—err, sleep. This time we begin by estimating the average sleep hours per night: T_{sleep}.

Although the general assumption is that we all need eight hours of sleep, many of us need less and some require more. While there are some humans who do not sleep, or have a different sleep structure, those cases are rare. I will use six hours as the minimum amount of sleep for some decent level of functioning, excluding medical students and residents. Many individuals feel best after nine hours of sleep, and some require ten. What do we get using these ranges?

$$T_{sleep} = \sqrt{6 \; hrs \times 10 \; hrs} = 7.7 \; hrs \approx 8 \; hrs \; per \; day$$

We are back to eight hours after all! Now we can estimate the average hours of sleep over a life span, using our previous life span estimate and time conversions:

$$Sleep_{time} = life_{duration} \times days_{year} \times hr_{day} \times T_{sleep}$$

Using our estimated values and canceling units, we have our results:

$$Sleep_{time} = \frac{77 \; \cancel{yrs}}{life} \times \frac{365 \; \cancel{days}}{\cancel{yr}} \times \frac{8 \; hrs}{\cancel{day}} \approx 225{,}000 \frac{sleep-hrs}{life}$$

Wow, this is significant! Since there are 365 × 24 hours in a year, this comes out to about 26 years. By the time we reach 77 years old, we have literally spent approximately one third of our life sleeping!

CHAPTER 14

WHAT IS THE DURATION OF "NOW"?

The famous Harvard psychologist and philosopher William James originated the concept of the "specious present": the sensation of an immediate brief span of time.[8] Is it possible to estimate what this span of time is? How would we go about it? Let's try to estimate this in two different ways.

First, we could estimate the time it takes to receive an incoming sensation and express it in awareness; that might be an upper limit on

---

[8] "We are constantly aware of a certain duration—the specious present—varying from a few seconds to probably not more than a minute, and this duration (with its content perceived as having one part earlier and another part later) is the original intuition of time." William James, *The Principles of Psychology* (New York: Henry Holt, 1890).

this time. So how long does it take to say "ouch"? As a neurologist, I had to measure nerve-conduction velocities as part of workups for neuromuscular syndromes. Knowing the typical sensory nerve-conduction velocity and the range of time necessary to express speech, we can derive an estimate.

Let's assume our subject is approximately six feet (two meters) in height (H). Now we need to estimate two quantities:

a. What is the sensory nerve-conduction velocity V_s?

b. What is the time to verbally respond to the ascending pain signal T_r?

With my experience in neurology, I know that sensory nerve-conduction velocities range from about 30–50 meters per second. You could also look this up in an electrophysiology textbook or on the Internet. Using my rough ranges, we can estimate an average velocity of:

$$V_s = \sqrt{30 \times 50} \; m/sec \approx 40 \; m/sec$$

What about speech responses? Neurologists know that short expletives can seem to be expressed almost immediately when surprised or startled. However, it does take some time to form the phonemic structure and issue the motor controls to create speech. I will estimate this as surely being greater than 1/100th of a second and perhaps less than half a second. Thus, we can estimate this as:

$$T_r = \sqrt{10 \times 500}\ msec \approx 70\ msec$$

Now we can estimate our "specious present" as the sum of the time required for the pain signal to be conducted to the brain and the time for a verbal response:

$$T_{present} = 70\ msec + \frac{2\ m}{40\ m/sec} \approx 120\ msec$$

This number is a little more than a tenth of a second.

Now let's take a different approach and see if we come up with estimates within a similar range. I suggest we use the well-known psychophysical phenomenon that humans do not see static motion when viewing successive images at a speed of approximately ≥ 20 frames per second (fps). This is how motion pictures work. Of course, now we use rapidly refreshed pixelated images on a screen instead of gelatin film projections. However, the human effect is indistinguishable: I see motion and not refreshed static screens.

The fact that we cannot see distinct static images presented to us at 20 fps suggests that we are not visually aware of changing events occurring within time intervals shorter than this. My hypothesis is that our sensation of "now" has to be less than or equal to the shortest duration of time that we can react to stimuli, whether visual, in the case of moving pictures, or pain, in the case of our first example. Here we can estimate this time interval as:

$$T_{present} = \frac{1\ frame}{20\ frame/sec} \approx 50\ msec$$

Now we have two estimates ranging from 50–120 millisecond (msec), but we still don't have an objective measure of "now" to compare to the figures. However, we do have modern estimates of reaction times. I looked at the Wikipedia page for mental chronometry, or reaction time, which stated that the elapsed time between a person being presented with a stimulus and initiating a motor response is about 200 msec.[9] Checking other websites, I found similar values ranging from the 200–300 msec, so our results are within a factor of 2 to 4x of these values. Not bad at all!

If our response time is approximately equivalent to our stimuli awareness time, we have at least a rough estimate for the upper end of the duration of "now." I say "the upper end" because most reaction times include the time for motor-response planning and execution, as in our first approach. It appears that our second and shorter estimate may be closer to the "now" moment because it is linked to the psychological experience of motion and does not have the added time for planning and executing motor responses.

As a final observation, the fact that there is a lag or duration to our "now" experience suggests two things. First, we may be too slow to sense or be aware of some rapid events that happen in the world around us. Second, since we are always operating with a time lag, we essentially are always reacting to the recent past. In other words, we actually don't experience true "now"!

---

[9] "Mental chronometry," Wikipedia, accessed February 26, 2020, https://en.wikipedia.org/wiki/Mental_chronometry.

CHAPTER 15

WHAT'S THE SURFACE AREA OF A HUMAN BEING?

Body surface area (BSA) is an important figure used for medication dosage, particularly for neoplastic syndromes. Given that the human body has a symmetric but complex body geometry, some people may find it surprising that BSA can be approximated with a simple geometric calculation. For our purposes, let's model the human simply as a rectangular solid of height H, width W, and thickness T. We know from simple geometry that the resulting area would be:

Rectangle solid area (ft^2) = $2[(T \times W) + (T \times H) + (W \times H)]$

Human size actually does not vary that much (you don't see adults who are two feet or twenty feet), so a reasonable range from four to eight feet covers everyone from pygmies to NBA basketball players. Now these are U.S. standard units; for scientific calculations, we want metric units (1 m ≈ 3 ft). Use the extremes to estimate H and convert to the MKS system:

$$H = \sqrt{4 \times 7} \approx 5 \, ft \times \frac{1 \, m}{3 \, ft} \approx 1.7 \, m$$

We'll use the same approach to estimate thickness and width. Assuming I am a typical human being, my thickness is probably between one and two feet, and my shoulder-to-shoulder width is more than two feet but less than three. Let's estimate these values and convert to MKS units:

$$T = \sqrt{1 \times 2} \approx 1 \, ft \times \frac{1 \, m}{3 \, ft} \approx 0.3 \, m$$

$$W = \sqrt{2 \times 3} \approx 2 \, ft \times \frac{1 \, m}{3 \, ft} \approx 0.7 \, m$$

For most of our calculations, we will not require more precision than this and will typically only want values to the nearest integer. Using our formula from the beginning of this chapter, we can now estimate the average body surface area as:

Body surface area $= 2 \times (0.3 \times .7 + .3 \times 1.7 + 0.7 \times 1.7) \, m^2 \approx 4 \, m^2$

So how did we do? Using Google, I found the "normal" body surface area:[10]

$$BSA = 1.7 \, m^2 \approx 2 \, m^2$$

Our estimate is roughly a factor of 2x the accepted value! This is a good approximation that illustrates how simple information can help appraise seemingly unknowable values.

---

[10] William C. Shiel Jr., "Medical Definition of Body Surface," MedicineNet, accessed February 26, 2020,
https://www.medicinenet.com/script/main/art.asp?articlekey=39851.

CHAPTER 16

WHAT'S THE VOLUME OF A HUMAN BEING?

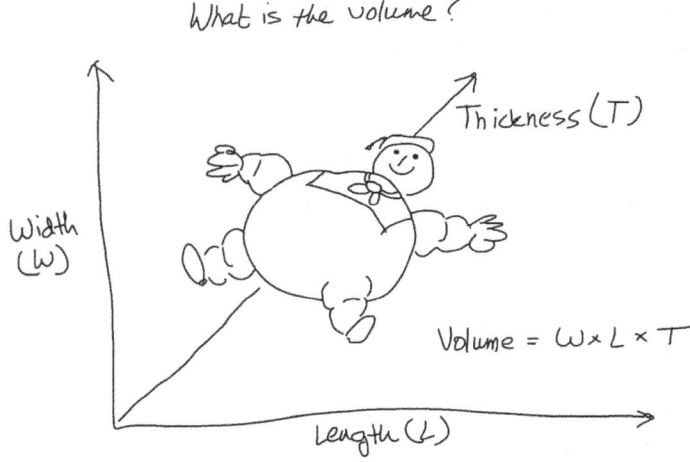

Now that we know how to estimate the surface area of a human being, let's go one step further and estimate our average volume. We'll again use our rectangular solid approximation for human anatomy. We know from simple geometry that the resulting area would be:

Rectangle volume (ft³) = H × T × W

Since we already have our MKS estimates for the three dimensions, we can immediately proceed to our calculation:

Body volume $= (1.7 \times 0.3 \times 0.7) \; m^3 \approx 0.3 \; m^3$

Going to the internet, I quickly found the average volume of a human body:[11]

$$Body\ volume = 66.4\ L\ or\ .066\ m^3$$

Here, our rough estimate has a factor of 5x difference, which again is less than the desired order of magnitude error. Let's look at another way to calculate body volume. We know the human body is mostly water, so we can estimate body volume using the standard weight of a human and the density of water.

Let's first estimate a standard body mass (M) for an adult human by taking the geometric average of plausible extremes—100 lbs (cachexia) to 300 lbs (morbid obesity)—and convert to MKS units:

$$M = \sqrt{100 \times 300} \approx 173\ \cancel{lbs} \times \frac{1\ kg}{2.2\ \cancel{lbs}} \approx 79\ kg$$

Knowing the body mass and water density allows us to estimate human body volume as:

$$Body\ volume = \frac{M}{D_w}$$

Using the density of water (1,000 kg/m³) as the density of a human being, we can now calculate the total volume as:

$$Body\ volume = 79\ kg \times \frac{m^3}{1000\ kg} \times \approx 8 \times 10^{-2}\ m^3$$

This is a much closer result with an error of less than 2x (19%, to be precise). Again, knowing a few things allows us to estimate other quantities that at first may seem unknowable.

---

[11] "Average Volume of Human Body," WolframAlpha.com, accessed February 26, 2020, https://www.wolframalpha.com/input/?i=normal+body+volume.

CHAPTER 17

HOW MANY CELLS IN A HUMAN BODY?

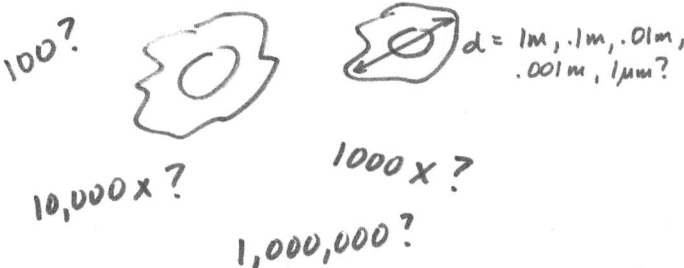

We are composite entities made up of cells. This is a reflection of scale from the physics level (quantum events to elementary particles to nucleons) to chemistry (atoms to molecules) to biology (cells to tissues to organs) to medicine (us). Much of human functioning and dysfunction (disease) arises at the cellular level. So how many cells make up the human body?

We now have a good estimate for the volume of the human body, so one way to estimate the number of cells is to simply divide the typical human body volume (V_h) by the average cellular volume (V_c). Of course, now the question is: what is the volume of a cell? Let's make some simple assumptions and see what we come up with. First, let's assume that cells can be approximated by small cubes of length L. Spherical or ellipsoidal shapes would probably be more realistic, but I want to keep the math simple.

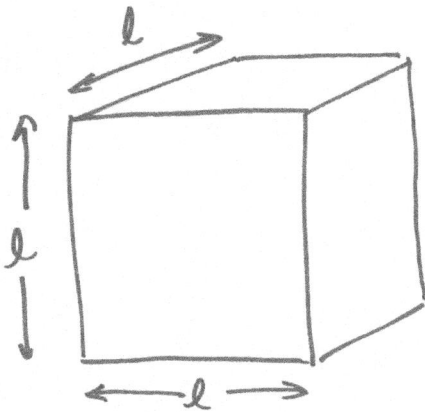

Figure 6. Cube volume $(m^3) = L^3$

Since we use microscopes to view the cells, we can assume they are larger than light and less than what the naked eye can resolve: between one micron and .1 millimeter, or 10^{-6} m to 10^{-4} m. We can estimate the length of our cellular cube as:

$$L = \sqrt{10^{-4}\ m \times 10^{-6}\ m} \approx 10 \times 10^{-6}\ m\ or\ 10\ \mu m$$

Using the simple cube volume, we can estimate the cellular volume (V_c) as:

$$V_c = (10 \times 10^{-6}\ m)^3 \approx 1{,}000 \times 10^{-18} m^3 = 10^{-15} m^3$$

Wow, that's small! Now let's use our calculated value for the volume of the human body based on the density of water and estimate the number of cells:

$$\#\ cells = N_c = \frac{8 \times 10^{-2}\ m^3}{1 \times 10^{-15}\ m^3} \approx 8 \times 10^{13}\ cells$$

So how did we do? A publication by Bianconi et al. in the *Annals of Human Biology* estimates the total cell number using bibliographical and mathematical approaches, based on a variety of organs and cell types, as: 3.72×10^{13} cells.[12] This differs from our estimated value only by a factor of 2x! Not bad for using simple geometry and algebra.

---

[12] Eva Bianconi et al., "An Estimation of the Number of Cells in the Human Body," *Annals of Human Biology* 40, no. 6 (2013): 463-471. https://www.tandfonline.com/doi/abs/10.3109/03014460.2013.807878.

CHAPTER 18

HOW MUCH ENERGY DO WE USE CLIMBING STAIRS?

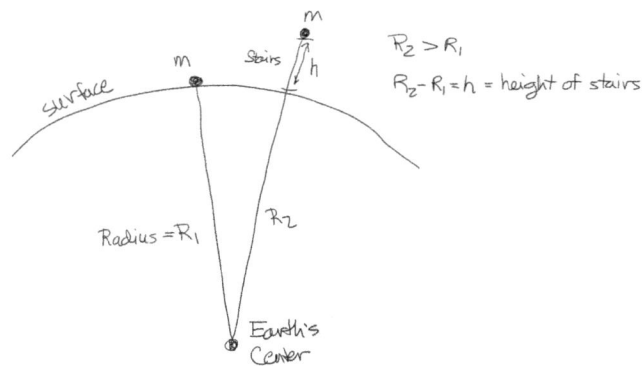

Once when I was eating lunch at a hospital in Las Cruces, New Mexico, a colleague wondered if we lost weight going up and down the stairs to see patients. One of my senior associates responded that any weight loss would be negligible; you'd be better off simply taking the elevator. This question intrigued me, and once I had more than a few hours of sleep, I estimated the answer. It's surprisingly straightforward and follows simple elementary physics.

How much energy do we use to climb a flight of stairs? Put another way, how much energy is required to raise a body of mass (M) up the height of the stairs? From the conservation of energy relation, this is simply the potential energy of the mass and the height:

Figure 7. Climbing stairs

We will use the classical expression for the potential energy of an object elevated to a height equaling the work done to lift it there:

$$E_p = mass \times height \times acceleration\ of\ gravity$$

At my current location, our office level is seven flights from the main floor. How high is a floor? In the U.S., most home and office ceilings are typically higher than eight or nine feet—let's say nine feet for an office building. Now we can estimate the height of one floor as:

$$H_{floor} = \sqrt{9\ ft \times 18\ ft} \approx 13\ ft$$

For seven floors, we can calculate the height and convert to metric units (1 m ≈ 3 ft):

$$H_{floor} = 13\ \frac{ft}{\cancel{floor}} \times 7\ \cancel{floor} \times \frac{1\ m}{3\ \cancel{ft}} \approx 30\ m$$

Next, we need to estimate the mass of our object. Assuming the object is me, and I am close to the average weight of a human male (about 70 kg or 154 lbs), we have our answer. What if I am heavier, perhaps 170 pounds? Then we convert my weight to kilograms using the following relationship:

$$M = 170 \, \cancel{lbs} \times \frac{1 \, kg}{2 \, \cancel{lbs}} \approx 85 \, kg$$

The result is actually closer to 80 kg, but I rounded the conversion factor of 2.2 lbs per kg to 2.0. For our purposes, this is close enough.

Finally, we need the gravitational acceleration at Earth's surface. We can obtain this from a standard physics text or the internet to find this is approximately 10 meters (m) per second squared. Now we can calculate the nominal potential energy in MKS units of joules (1 joule = 1 kg-m/s2; pronounced like "jewels") for climbing seven flights of stairs:

$$E_p = 85 \, kg \times 30 \, m \times \frac{10 \, m}{s^2} \approx 25{,}500 \frac{kg \, m}{s^2} \approx 26 \, kJ$$

Before we convert to units more familiar to those with a biological bias, we should realize that biological entities are not able to convert food energy to work (walking upstairs) with complete efficiency. The efficiency of Carnot engines (think of your car) probably approaches some of the highest levels in most mechanical systems at around 30–40%. My recollection from biochemistry is that living systems operate at similar levels, so let's assume an average efficiency of 35% (ε) in converting food energy to mechanical work. We can state the total food energy used in climbing our stairs as:

$$Total \, food \, energy = \frac{E_p}{\varepsilon} \approx \frac{26 \, kJ}{0.35} \approx 74 \, kJ$$

Is this a significant amount of energy? Let's convert to more conventional units to make it clearer, using the convention that one food-value calorie = 1,000 "thermal" calories.[13]

$$Total \, food \, energy = 74 \times 10^3 \, J \times 1 \frac{cal}{4 \, \cancel{J}} \approx 19 \, kc \, or \, 19 \, Calories$$

Ok. It's not a large amount, but it certainly is not negligible. If you

---

[13] Hugh D. Young and Roger A. Freedman, *Sears & Zemansky's University Physics with Modern Physics*, 14th ed. (London: Pearson, 2015).

go up and down the stairs five to ten times a day, this adds up to significant food calories burned. So it's a smart idea to avoid the elevator!

How does this figure compare to values from the public domain? I went to livestrong.com[14] as a simple check and found an estimate of five to ten calories burned per flight of stairs. Assuming that our definitions of "flights" and body mass are similar, my result of 19 calories is only 2x to 3x different from the website's range of 35–70 calories estimated for a seven-flight climb. This is well within the 10x Fermi estimate range and in fact is quite close!

---

[14] Karen Spaeder, "Calories Burned Climbing One Flight of Stairs," Livestrong.com, last modified June 26, 2019, https://www.livestrong.com/article/301539-calories-burned-climbing-one-flight-of-stairs/.

CHAPTER 19

WHAT IS THE METABOLIC RATE OF A HUMAN BEING?

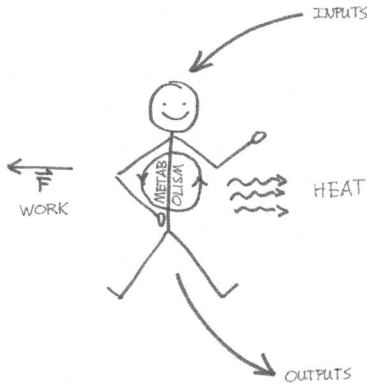

B elieve it or not, we are heat engines. Surprised by that? Well, think about it. We continuously ingest material, process it, and produce waste byproducts such as heat. As a consequence, we are warm. This may sound silly to point out, but most of us don't fully appreciate the clear mechanical aspects of our bodies. You'll notice that I use heat and energy interchangeably in this chapter; I do this because heat and energy are equivalent from a physics perspective.

How much energy do we need to function at a basic level? Let's approach this question in two separate ways. (Note: This is a favorite problem in many books on estimation and physics, so my approach is not unique. I encourage you to refer to other authors' solutions to this problem.)

First, let's begin by defining the basal metabolic rate (BMR) in terms of the amount of food we eat over time:

$$BMR = power = \frac{energy\ intake\ (E)}{time\ (T)}$$

In the U.S., the typical range of energy intake (food) goes from about 1,800 cal/day up to 3,500 cal/day or more. (Of course, smaller individuals and dieters may consume less than 1,800 calories, and some large individuals and athletes may consume much more than 3,500.) Using the average range, let's calculate the geometric mean:

$$E = \sqrt{1{,}800\frac{cal}{day} \times 3{,}500\frac{cal}{day}} \approx 2{,}500\frac{cal}{day}$$

Now we need to convert this estimate into MKS units for power, noting that a food calorie is approximately 4,000 joules. Power is defined as the rate of energy processed in joules per second and is measured in watts:

$$BMR = 2{,}500\frac{\cancel{cal}}{\cancel{day}} \times \frac{1{,}000\ \cancel{cal}}{\cancel{cal}} \times \frac{4\ J}{\cancel{cal}} \times \frac{1\ \cancel{day}}{24\ \cancel{hrs}} \times \frac{1\ \cancel{hr}}{3{,}600\ sec}$$

$$\approx 120\ watts$$

On average, humans use about 100 watts a day. Is there another way to check this result? Yes. We can use a relationship from thermal physics known as the Stefan-Boltzmann law (SBL), which states that "black bodies" radiate energy H (heat, measured in watts) at a power level/unit area proportional to the fourth power of the absolute temperature in kelvins, or:

$$H = Ae\sigma(T_B^4 - T_S^4)$$

In this version of the equation, (A) is the area of a human being in meters, (e) is a unitless quantity called the emissivity, which for human skin approximately equals 1, sigma (σ) is the SB constant, and (T_B) and

(T_s) are the body surface and surrounding temperature values in kelvins. Finally, the way I have written the equation subtracts the surrounding temperature from the body temperature. This allows us to estimate the net heat (energy) transferred by the human body, accounting for the heat absorbed by the surroundings.

So, what do we do next? First, we will need to convert the common temperatures we live with into units of Celsius and then kelvins. This may seem cumbersome, but in thermal physics it is important to convert and use the absolute temperature scale of kelvins. Let's start by considering skin temperature, which is often a few degrees lower than our core 98.8° temperature on the Fahrenheit scale. Assuming it is about five to ten degrees cooler than core temperature, we can determine average skin temperature by combining the Fahrenheit to Celsius conversion with the conversion to kelvins. This gives us the value for the human skin as:

$$T_{skin} = \frac{5}{9} \times (91 - 32) = 33\ C + 273\ K = 306\ K$$

Next we calculate the surround absolute temperature. If our surroundings are 70°F on a pleasant spring day, we use this equation:

$$T_{surround} = \frac{5}{9} \times (70 - 32) = 21\ C + 273\ K = 294\ K$$

Knowing that the emissivity from human skin is about 1, and the human body surface area from our previous calculations is about 2 m², we can obtain the SB constant from a textbook or the Web and then solve for the net radiative heat (energy) flow:

$$H = 2\ m^2 \times 1 \times \left(6 \times 10^{-8} \frac{W}{m^2 K^4}\right)(306\ K^4 - 294\ K^4) \approx 150\ W$$

As you can see, this result is very close to our first estimate of 120 watts. However, we have derived it using the completely different approach of thermal physics. Not bad!

CHAPTER 20

HOW MANY WORDS IN A BOOK?

This may seem an odd question to ask when we're discussing healthcare, but it will make sense as we go through this and other related problems. Language is the primary tool we use to convey information, particularly if is based on written text. In the healthcare field, books form the backbone of medical knowledge and contain information that students and practitioners must learn and maintain as part of their practice.

In order to develop a feeling for the amount of information our healthcare providers must absorb, it is useful to try to quantify this. Let's begin by asking, "How many words are in a typical line of English text?" The shortest lines I've read in a book are five words. It's improbable that one line could contain more than 50 words, even if the type size is small, so we can use the geometric mean to estimate the average as:

$$Words_{line} = \sqrt{5 \times 50} \approx 16 \; words \; per \; line$$

Now we ask, "How many lines are on a typical page of English text?" In my reading experience, 20 lines per page would be short and 50 lines would be on the long side. Our geometric estimate thus becomes:

$$Lines_{page} = \sqrt{25 \times 50} \approx 35 \; lines \; per \; page$$

Finally, we need to estimate the number of pages per book. Doing a quick review of my bookshelf at home, I found that my books appear to range between about 100+ pages to about 500 for longer books. I will use these numbers as a reasonable range:

$$Pages_{book} = \sqrt{100 \times 500} \approx 225 \; pages \; per \; book$$

Now we can estimate the average number of words in a book, using our estimates:

$$N_{words} = words_{line} \times lines_{page} \times pages_{book}$$

Or using our estimated values and canceling units, we have our results:

$$N_{words} = \frac{16 \; words}{\cancel{line}} \times \frac{35 \; \cancel{lines}}{\cancel{page}} \times \frac{225 \; \cancel{pages}}{book} \approx 1 \times 10^5 \frac{words}{book}$$

According to Writer's Digest, 80,000–90,000 words is a good range for an adult novel.[15] This figure is practically spot-on and is well within our factor of 10x, so this is a good estimate!

---

[15] Chuck Sambuchino, "Word Count for Novels and Children's Books: The Definitive Post," *Writer's Digest,* October 24, 2016, https://www.writersdigest.com/guest-columns/word-count-for-novels-and-childrens-books-the-definitive-post.

CHAPTER 21

HOW MUCH MEMORY STORAGE DOES A BOOK REQUIRE?

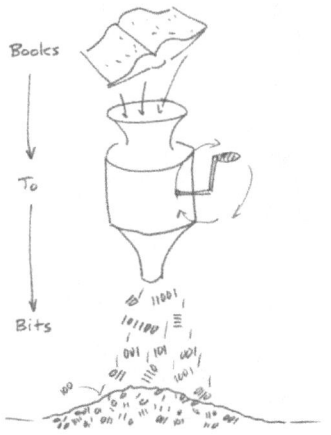

In our previous chapter we estimated the number of words in a book. That may seem like a trivial exercise, but it becomes important when we want to know how to store and share information in a book or any other media. Specifically, we can build on our previous work to estimate how much digital memory storage is required for an average book.

To do this, I need to introduce the concept of a binary digit, or "bit." This is a two-state number (0 or 1) that is the fundamental unit of binary arithmetic underlying our modern digital computers. Based on recent advances in classical and quantum information theory, the "bit" can be considered the fundamental unit of information that gives rise to all of physics and our universe.[16] That's beyond our discussion here, however, so let's return to the task at hand.

To estimate the total storage in bytes (1 byte = 8 bits), we need to estimate the number of characters required per word and use that with the average words per book. Assuming we are using English texts and symbols, the common data standard is one byte per character. Let's begin by asking, "How many characters are in a typical word of English text?" Native English speakers know that the shortest word is one character (there are three: "a," "I," and "O") and the longest words are rarely more than 15 characters. So let's again use the geometric mean to estimate the average:

> The longest word in the English language, the chemical name for the titin protein found in humans, has 189,819 letters. Several other scientific words have between 30–45 letters.
> "Supercalifragilisticexpialidocious" has 34 letters, "antidisestablishmentarianism" has 28, and "incomprehensibilities," weighing in at 21 letters, set the record in the 1990s as the longest word in common usage.
> Shundalyn Allen, "14 of the Longest Words in English," Grammarly, accessed February 26, 2020, https://www.grammarly.com/blog/14-of-the-longest-words-in-english/.

$$Characters_{word} = \sqrt{1 \times 15} \approx 4 \ characters \ per \ word$$

We have already gone through the exercise of estimating the number of words in a book, so we can use this figure with our geometric character estimate and the ASCII bytes per character to yield an answer for the number of characters:

$$N_{characters} = \frac{4 \ \cancel{characters}}{\cancel{word}} \times \frac{10^5 \ \cancel{word}}{book} \times \frac{1 byte}{\cancel{character}}$$

---

[16] John Archibald Wheeler, "Information, Physics, Quantum: The Search for Links," *Proceedings of the 3rd International Symposium of Quantum Mechanics in the Light of New Technology* (Physical Society of Japan, Tokyo, 1990).

$$\approx 4 \times 10^5 \frac{bytes}{book} \approx 400 \frac{kbytes}{book}$$

The average file size and average number of pages of a Kindle eBook is 2.6 MB and 300 pages, respectively.[17] We estimated .4 MB (400 KB) and 225 pages; our figures are within a factor 6x and 1x of the website results. Given that we had not included any graphics in our book estimate, it is not surprising that we were off by a factor of six, as graphic objects require greater digital memory than text. Still, both estimates are well within the 10x error we should expect with the Fermi approach.

---

[17] "The Average Size of a Kindle eBook," Elite Authors, January 17, 2019, https://eliteauthors.com/blog/the-average-size-of-a-kindle-e-book/.

CHAPTER 22

HOW MANY BOOKS IN THE LIBRARY OF CONGRESS?

Let's use our previous estimates to predict an interesting, verifiable question: How many books are in the Library of Congress (LOC)? If you've never visited the LOC, it's worth a trip to Washington, D.C. This is a premier library with some beautiful art and historical documents.

Let's proceed with the strategy of first determining the volume of the physical library and then dividing this by the typical book volume to estimate the total number of books. From my few trips to the LOC, I recall it being about a city block in area and about five stories in depth. From memory, I recall that a city block is approximately a tenth of a mile in length. Using this assumption, we can estimate the area using the mile-to-feet conversion:

$$Block_L = 1 \; \cancel{block\;length} \times \frac{0.1 \; \cancel{mile}}{\cancel{block\;length}} \times \frac{5{,}280 \; ft}{\cancel{mile}} \approx 528 \; ft$$

From this we can determine the area of the block in feet, assuming the block is square:

$$Block_{area} = 528 \; ft \times 528 \; ft \approx 279{,}000 \; ft^2$$

Next, we determine the thickness of the building. Let's assume the floor of a building in the U.S. is probably between 9–14 feet. We'll estimate the floor height as:

$$Floor_H = \sqrt{9 \times 14} \approx 11 \; feet \; per \; floor$$

Now we can estimate the total volume as a five-story, squared block building. This becomes:

$$LOC_{vol} = 279{,}000 \; ft^2 \times 5 \; \cancel{floors} \times \frac{11 \; ft}{\cancel{floor}} \approx 15 \times 10^6 \; ft^3$$

Although we now have an overall estimated volume for the LOC, we need to realize that the entire building cannot be filled with books. There must be room for the floors, desks, doors, administrative offices, walkways, restrooms, and other necessary spaces. Let's estimate a percentage for what I will term "non-storage space." I would be surprised if more than 50% of the LOC comprised such space. Also, it seems implausible that this volume should be less than about 10%. Therefore, I will assume that non-storage space is between these two extremes, which I will estimate with our standard geometric mean:

$$Nonstorage \; space = \sqrt{10\% \times 50\%} \approx 22\%$$

Now we can adjust our estimate for the LOC volume to remove the space that cannot be used for books:

$$LOC_{Vadj} = 15 \times 10^6 \; ft^3 \times (1 - 0.22) \approx 12 \times 10^6 \; ft^3$$

At this point we have our adjusted LOC volume; now we need to estimate a typical book size. The range of book sizes goes from small paperbacks up to oversize coffee-table books. It's difficult to simply assume a standard volume, so I will use our Fermi ranges for typical length, width, and height of books to estimate a figure. I will start with book length on a shelf and assume this will range from 0.5–1.5 feet in length. For height, I will again assume a range from 0.5–1.5 feet. For width, I will assume a minimum of 1 inch (1/12 ft) up to 8 inches (0.75 ft). The geometric estimates become:

$$Book_L = \sqrt{0.5\ ft \times 1.5\ ft} \approx 0.9\ ft$$

$$Book_H = \sqrt{0.5\ ft \times 1.5\ ft} \approx 0.9\ ft$$

$$Book_W = \sqrt{0.08\ ft \times 0.75\ ft} \approx 0.3\ ft$$

Our typical book volume then becomes:

$$Book_V = 0.9\ ft \times 0.9\ ft \times 0.3\ ft \approx 2 \times 10^{-1}\ ft^3$$

Finally, we are able to estimate the total number of books in the LOC as follows:

$$N_{books} = 12 \times 10^6\ ft^3 \times \frac{1\ book}{2 \times 10^{-1}\ ft^3} \approx 60 \times 10^6\ books$$

Consulting the Library of Congress website reveals that its collection as of the end of fiscal year 2018 numbered 168 million items, including more than 39 million cataloged books and other print material.[18] This puts our estimate within less than a factor of 2x the actual value. This is quite good!

To go a little further and link this analysis with our previous chapter evaluating the digital storage required to comprise a book, let's estimate the digital memory needed to contain the LOC's collection of books. We estimated about 0.5 MB per book and our web reference had a value of

---

[18] "Year 2018 at a Glance," Library of Congress website, accessed February 26, 2020, https://www.loc.gov/about/general-information/#year-at-a-glance.

2.5 MB. If we use these values as our range, the geometric mean size is:

$$Book_L = \sqrt{0.5\ MB \times 2.5\ MB} \approx 1\ MB$$

We can now calculate that about 40 million MB (1 MB per book) of digital memory storage is required to contain the LOC's collection of books. As written, this is a cumbersome number. Let's convert it to terabytes (TB) as follows:

$$40\ million\ MB = (40 \times 10^6) \times (10^6)\ bytes = 40 \times 10^{12}\ bytes = 40\ TB$$

According to market intelligence company IDC, the "global datasphere" in 2018 reached 18 zettabytes.[19] One billion terabytes equals one zettabyte. Thus, the total Library of Congress collection accounts for only 10^{-7}% of the world's estimated data!

The point of this exercise is to put into context the staggering amount of information that humans are exposed to. Consider the medical information that a healthcare provider had to know 50–100 years ago and compare that to the information providers need today. While our technology has scaled nonlinearly and we can access information on a mobile phone that was unthinkable just a few decades ago, our minds and ability to assimilate this information have not kept pace.

---

[19] "How Much Data Is There in the World?," Bernard Marr website, accessed February 27, 2020, https://www.bernardmarr.com/default.asp?contentID=1846.

CHAPTER 23

WHAT IS THE SPEED OF READING?

We have spent several chapters estimating the size of books and how much memory storage they require. Now let's estimate how quickly someone can read a book, journal article, or blog post. This is an important determination; reading a paper on a new treatment for Alzheimer's or laboring through standard operational procedure (SOP) training requires an investment of time that could be spent on other essential activities. Let's begin by estimating the speed of reading and then examine some of the implications to healthcare practitioners.

How long does it take to read an average book? From our previous examples, we have developed estimates for the length of books, so let's estimate the time it takes to read a page and then build up an estimate for an entire book. Using myself as an example, I think it typically takes me a few minutes to go through an average page of text from an average novel or book of minimal complexity. I estimate that it rarely takes as much as five minutes and typically does not take less than two minutes. Therefore, I can estimate my easy reading time as:

$$R_t = \sqrt{2 \times 5} \approx 3 \; mins \; per \; page$$

Now that estimation was for easy reading and may be too conservative. However, I think it is intuitively obvious that reading a page of dense mathematics, charts, and numbers is quite a bit more intellectually challenging than perusing a page from a detective novel. Let's evaluate how long it might take to read a page from a more complicated text such as a medical or biostatistical journal.

I will estimate my typical reading time to be somewhere between my easy-reading average and ten minutes per page, in which case I would be reading with a notepad to follow a derivation or mathematical discussion. This gives me a notably more conservative reading-speed estimate for dense texts:

$$R_t = \sqrt{3 \times 10} \approx 5 \; mins \; per \; page$$

What does this tell us? From my background knowledge, I know that good typists generally produce text on the order of 100 words per minute (I'm dating myself here, as the word "typist" is not used much anymore). I would expect that typical reading speeds would certainly be faster than the standard typing speed. From our previous estimates for books, a typical line of text is 16 words, with 32 lines of text per page and 225 pages per book. Using our values for a page of text, I can estimate my reading speeds for "easy" and "technical" books as:

$$R_{easy} = \frac{16 \; words}{\cancel{line}} \times \frac{32 \; \cancel{lines}}{\cancel{page}} \times \frac{1 \; \cancel{page}}{3 \; mins} \approx 170 \frac{words}{min}$$

$$R_{tech} = \frac{16 \, words}{\cancel{line}} \times \frac{32 \, \cancel{lines}}{\cancel{page}} \times \frac{1 \, \cancel{page}}{5 \, mins} \approx 125 \frac{words}{min}$$

The easy reading is certainly faster than most typing speeds I'm aware of. The speed for reading technical material seems a bit low to me, but this may be biased by my own practice of reading more physics and mathematically dense material. However, assuming I am a reasonable example of a healthcare provider, it would seem sensible that a typical reading speed for most material lies between these two boundaries.

To check my accuracy, I found estimates for reading speeds from several websites. According to some theorists, the average adult reading speed is 250 words per minute, and the average college student reading speed is 300 words per minute. Technical material is read at a much slower rate, about 50–75 words a minute. So in summary, based on results from speed-reading tests, my estimates are a little low for average reading, and off by less than a factor of 2x for the technical material.

Why is this important for thinking about healthcare and medicine? It comes down to time. A recent paper by Silverberg and Ray in the *Journal of General Internal Medicine* evaluated the word counts of 715 articles in five of the top medical journals. The range for the main text varied from 2,700 words (*New England Journal of Medicine*) to 4,400 words (*British Medical Journal*).[20] Using the geometric mean estimate for average journal word count, we obtain:

$$N_J = \sqrt{2{,}700 \times 4{,}400} \approx 3{,}500 \, words \, per \, article$$

Assuming that my technical reading speed represents an optimistic rate, we can calculate the time to read and digest a typical medical journal article:

$$R_{time} = \frac{3{,}500 \, \cancel{words}}{article} \times \frac{1 \, min}{125 \, \cancel{words}} \approx 30 \, mins \, per \, article$$

---

[20] Orli Silverberg and Joel G. Ray, "Variations in Instructed vs. Published Word Counts in Top Five Medical Journals," *Journal of General Internal Medicine* 33, no. 1 (January 2018): 16-18, https://www.ncbi.nlm.nih.gov/pmc/articles/PMC5756173/.

Many healthcare providers are required to read and test on journal articles, exams, etc., over the course of a year to maintain license certification. This is time spent on top of patient care and documentation. If we assume that a typical practitioner may read up to five articles per week for fun and/or certification, this can easily add up over a year's time:

$$R_{time} = \frac{30 \ mins}{\sout{article}} \times \frac{5 \ \sout{articles}}{week} \times \frac{52 \ \sout{weeks}}{year}$$

$$\approx 7{,}100 \ mins \ or \ 120 \ hours \ per \ year$$

Given that a work month is about 160 hours, this is a significant time investment. With the number of new medical journals, research, and publications increasing dramatically, finding the time to keep up with the latest literature in one's field becomes a true challenge.

CHAPTER 24

MY HOUSE ROCKS!

How many of you worry about meteorite impacts? Do any of you think you need a special insurance policy for your home for cosmic catastrophes? While it may sound like a joke, and the risk is indeed small, there have actually been cases of homes damaged by small meteorite impacts. One famous case occurred in 1954, when Ann Hodges of Alabama was struck by a softball-sized black rock that crashed through her ceiling.[21] So it does occur, but it's very rare.

---

[21] Justin Noble, "The True Story of History's Only Known Meteorite Victim," *National Geographic* website, February 20, 2013, updated November 30, 2016, https://www.nationalgeographic.com/news/2013/2/130220-russia-meteorite-ann-hodges-science-space-hit/.

How rare is it? Let's find out by thinking through the problem. If we know how often meteorites fall to earth and make it through the atmosphere, we should be able to make a reasonable Fermi-style estimate. Let's pose the question this way: What is the probability that your house will be struck by a meteorite in one year?

First, let's determine how many meteorites hit the earth in a year. A radar astronomist at the Smithsonian estimated that between 18,000 to 84,000 meteorites larger than 10 grams fell to Earth each year.[22] We can estimate the strike rate:

$$SR_m \approx \sqrt{18{,}000 \times 84{,}000} \approx 39{,}000 \; hits/year$$

Assuming that most of these burn up in the atmosphere (let's say 75%), then the likely number would become:

$$SR_m \approx (1 - .75) \times 39{,}000 \frac{hits}{year} = 9{,}800$$

Next, let's assume that the meteorite hits are uniformly distributed over the earth. This implies that the vast majority will actually land in the oceans or other uninhabited areas, which makes sense. If we know the area of the earth, average home size, and the number of homes (one, for your single home), then we can write the probability of a house strike in one year as:

$$P(hit) = \frac{A_H \times N_H \times SR_m}{A_E}$$

The size of American homes typically ranges between 1,200 ft² to 8,000 ft². Let's estimate the geometric-mean home area as:

$$A_H \approx \sqrt{1{,}200 \; ft^2 \times 8{,}000 \; ft^2} \approx 3{,}100 \; ft^2$$

---

[22] Lynn Carter, "How Many Meteorites Hit Earth Each Year?," Ask an Astronomer at Cornell University, February 9, 2003, last modified July 18, 2015, http://132.236.6.82/the-universe/cosmology-and-the-big-bang/75-our-solar-system/comets-meteors-and-asteroids/meteorites/313-how-many-meteorites-hit-earth-each-year-intermediate.

Dealing with imperial units is annoying, so let's convert to MKS units by assuming 1 m ≈ 3 ft, so 3,100 ft² becomes 344 m². Referring to my old physics textbook appendix with useful constants,[23] the area of the earth is approximately 5×10^{14} m².

Now we can use this information to estimate the probability of your home being struck by a meteorite in any one year:

$$P(hit) = \frac{\frac{344 \; \cancel{m^2}}{\cancel{house}} \times 1 \; \cancel{house} \times \frac{9{,}800 \; hits}{\cancel{year}} \times 1 \; \cancel{year}}{5 \times 10^{14} \; \cancel{m^2}} = 7 \times 10^{-9}$$

This is a very small number. What would the answer be if we ask the same question about all the homes in the U.S. during any given year? The result would be the same for the one home multiplied by the total number of homes. If we assume there is a home for every four people (parents with two children), then the number of homes could be estimated as:

$$N_H = \frac{US_{pop}}{pop/home} = \frac{300 \times 10^6 \; \cancel{pop}}{4 \; \cancel{pop}/home} \approx 75 \times 10^6 \; homes$$

This seems like a larger number than I would expect, but hopefully within an order of magnitude of the correct answer. Now we simply multiple the single home probability x the number of homes:

$$P(hit) = \frac{7 \times 10^{-9}/year}{\cancel{home}} \times 75 \times 10^6 \; \cancel{homes} = 0.5/year$$

This is interesting. Simply due to the large number of homes and their associated areas, it appears there's about a 50% chance that at least one U.S. home will be struck in any given year. This has certainly occurred in the past, and the rare occurrence always makes the news. Let's hope you aren't the next meteor celebrity!

---

[23] The first edition of *University Physics* was published by Francis Sears and Mark Zemansky in 1949. The popular physics textbook is now in its fourteenth edition: Hugh D. Young and Roger A. Freedman, *Sears and Zemansky's University Physics with Modern Physics*, 14th ed. (London: Pearson, 2015).

CHAPTER 25

STRUCK BY LIGHTNING

During my practice as a neurologist, I was fascinated by cases of patients who had suffered injuries from lightning. Having been close to lightning while rock climbing years ago, I still have a vivid memory of the ozone smell and distinct tingling feeling of being near a strike. Fortunately for my climbing team, everything turned out fine. However, many people are struck by lightning each year, and some die as a result.

Let's estimate the probability and number of injuries due to lightning in the U.S. every year. First, let's assume that an individual is injured if they are within one meter of a strike. The hazard area of a strike is:

$$A_L = \pi r^2 \approx 3 \times (1m)^2 \approx 3\ m^2$$

Like our estimate of the probability of a meteorite hitting a house, let's assume that lightning strikes occur uniformly over the earth at an average rate of 30 strikes per second. First, let's determine how many lightning strikes hit the earth in a year. Now we can estimate the strike rate:

$$SR_L = \frac{30\ strikes}{\sec} \times \frac{60\ \cancel{secs}}{min} \times \frac{60\ \cancel{mins}}{hr} \times \frac{24\ \cancel{hrs}}{day} \times \frac{365\ \cancel{days}}{yr}$$

$$\approx 10^9 \frac{strikes}{yr}$$

Since we are interested in the number of strikes per year in the U.S. and the resulting number of injuries, we need to modify our strike rate to reflect activity over the continental United States. Rather than look up the area of the country, let's approximate it by a rectangle of approximately 3,000 miles wide (three time zones at about 1,000 miles each) and 2,000 miles high, for a total of approximately:

$$A_{US} \approx 2,000\ mi \times 3,000\ mi \approx 6 \times 10^6\ \cancel{mi^2} \times \frac{(1,600\ m)^2}{\cancel{mi^2}}$$

$$\approx 1 \times 10^{13}\ m^2$$

Now we need to estimate the area of the earth and use that information to estimate the strike rate for the U.S. To make things simple, we will approximate the earth as a cube of 24,000 miles (24 time zones) in circumference. Each cube face can be estimated as one fourth of the circumference, or 6,000 miles. We then approximate the total area as the sum of the areas of the six faces:

$$A_E \approx ((6,000\ mi \times 6,000\ mi) \times 6) \times \frac{(1,600\ m)^2}{\cancel{mi^2}} \approx 5 \times 10^{14}\ m^2$$

Now we can update our estimated strike rate to reflect the uniformly distributed number over the continent over one year:

$$SR_{LUS} = \frac{A_{US}}{A_E} \times SR_L \approx \left(\frac{1 \times 10^{13} \, \cancel{m^2}}{5 \times 10^{14} \, \cancel{m^2}}\right) \times \left(10^9 \frac{strikes}{year}\right)$$

$$\approx 10^7 \frac{strikes}{year}$$

If we assume that the U.S. population is uniformly distributed across the continent (obviously a very simplistic assumption), and we know the area of the continent, average strike area, and the number of individuals in the U.S., then we can write the probability of a strike injury for an individual in one year as:

$$P(hit) = \frac{A_L}{A_{US}} \times SR_L \times exposure_t$$

With an exposure time of one year, this becomes:

$$P(hit) = \frac{3 \, \cancel{m^2} \times \frac{10^7 \, strikes}{\cancel{year}} \times 1 \, \cancel{year}}{1 \times 10^{13} \, \cancel{m^2}} = 3 \times 10^{-6}$$

This is a small probability for any one individual. However, if we ask how many individuals are injured over the course of a year, we can estimate this as the product of the individual probability and the total U.S. population:

$$N_L \approx N_{US} \times p(hit) \approx (300 \times 10^6) \times (3 \times 10^{-6})$$
$$\approx 900 \, individuals$$

So how good is our estimate? I went to the Web to find the most current statistics on lightning. According to the National Oceanic and Atmospheric Administration (NOAA), based on data collection from 2009 through 2019, the U.S. averages about 27 deaths annually due to lightning injuries.[24]

The average mortality from a lightning strike is quoted as approximately 10%. From my results, this implies that the average is approximately 90 deaths (the 10% mortality rate × 900) per year. We're off from

---

[24] "National Weather Service Lightning Fatalities in 2019: 20," National Weather Service website, accessed February 27, 2020, https://www.weather.gov/safety/lightning-fatalities.

the official statistics by a factor of only 3x. This is quite good, given the crude approximations we used. Depending on your assumptions and approximations, your resulting estimate could differ by up to 10x (the standard Fermi goal) or more.

CHAPTER 26

HOW MANY DOCTORS ARE THERE IN THE UNITED STATES?

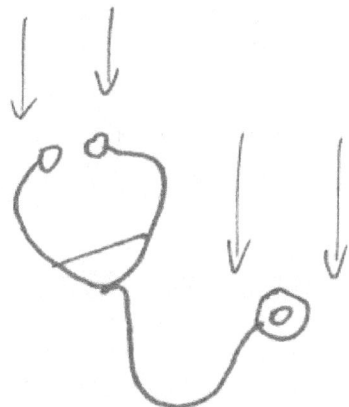

To answer this question, we need to estimate several pieces of information:

1. How often does the average person see a doctor in the United States? Call this visit frequency V_f.
2. How much time does the average doctor visit take? Call this visit duration V_t.
3. How many visits can the average doctor have in a given period of time? Visit volume V_v.

Let's assume the question refers only to outpatients in one year. This is a reasonable assumption, since hospitalization is generally rare compared to other reasons for doctor visits or interactions. For example, I generally visit one of my colleagues between one and three times in a year. However, some individuals go to the doctor more frequently. To ensure that the estimate is representative across a population that likely includes sicker individuals, let's estimate the upper boundary as 12 visits per year. Using the geometric mean formula, we can estimate the overall average as:

$$Visit_{freq} = \sqrt{1 \times 12} \approx 3 \text{ visits per year}$$

How much time does the average doctor visit take? As a former practicing physician, I find it difficult to imagine follow-ups that take less than five minutes, but have definitely seen office visits that can go an hour or more. For convenience, I will estimate the visit duration is between 5–60 minutes:

$$Visit_{time} = \sqrt{5 \times 60} \approx 17 \text{ mins per visit}$$

Finally, how many visits can a doctor have in a year? Let's assume that a typical working day is between 8–10 hours, giving a geometric mean estimate:

$$Work_{day} = \sqrt{8 \times 10} \approx 9 \text{ working hours per day}$$

Assuming that we want our doctors to at least be able to eat and have breaks, let's subtract one hour and make this a standard eight-hour working day. If we use the standard working time for a year ($work_{year}$ = 2,000 hours) and the minutes-to-hours conversion factor (min_{hr} = 60 mins/hr), we can estimate the average number of visits that a working physician can accommodate in a year as follows:

$$N_{visit} = work_{year} \times min_{hr} \times \frac{1}{visit_{time}}$$

or

$$N_{visit} = \frac{2{,}000 \; \cancel{hrs}}{yr} \times \frac{60 \; \cancel{mins}}{\cancel{hr}} \times \frac{1 \; visit}{17 \; \cancel{mins}}$$

$$\approx 7 \times 10^3 \frac{visits}{yr} \; per \; doctor$$

Finally, we are now in a position to answer the original question regarding the number of doctors in the U.S. We can assume that the number of doctors is proportional to the number of possible visits and the overall U.S. population, and inversely proportional to the total annual visits per doctor:

$$N_{doctor} = US_{pop} \times visit_{freq} \times \frac{1}{N_{visit}}$$

Using our estimated values and canceling units, we have our results:

$$N_{doctor} = 300 \times 10^6 \; \cancel{pop} \times \frac{3 \; \cancel{visits}}{\cancel{yr/pop}} \times \frac{1 \; doctor}{7 \times 10^3 \; \cancel{visits}}$$

$$\approx 10^5 \; doctors$$

Thus, we estimate on the order of 10^5 (100,000) total doctors in the U.S. The most recent figures put the number of active patient care MDs in the U.S. as about one million doctors.[25] We are within an order of magnitude, so not bad!

---

[25] "Professionally Active Physicians," State Health Facts, Kaiser Family Foundation website, March 2019, https://www.kff.org/other/state-indicator/total-active-physicians/?currentTimeframe=0&sortModel=%7B%22colId%22:%22Location%22,%22sort%22:%22asc%22%7D.

CHAPTER 27

HOW MANY PATIENTS CAN A DOCTOR SEE?

How busy is your primary care doctor? My guess is quite busy, but why guess when we can estimate it? I will do this calculation a little more carefully than some of my earlier ones to illustrate some different aspects of a physician's life. First, we will make some basic assumptions about the overall work timing and operations of a doctor's office.

Let's start by assuming that our doctor gets two weeks off a year (some get none). During a typical workday, a doctor has a half-hour lunch and 15-minute bio-break, or 0.75 hours a day of down time, plus at least an hour of documentation and administration time per day. This makes a total of about

1.75 hours per day when the doctor is not with a patient. In the spirit of approximation, let's make this two hours, in case we're underestimating the time documenting charts, returning calls, ordering tests, etc.

Patients can be new patients, current patients with new conditions, or follow-ups; each kind requires a different amount of time for evaluation. Rather than get extremely detailed, we'll estimate the overall average time to see follow-up patients as about 10 minutes, and new patients require about 20 minutes or so. Of course, if we were modeling a situation with sicker patients, these average numbers could be much larger. Our simple model, however, becomes:

$$t_{eval} = \sqrt{10 \; mins/pt \times 20 \; mins/pt} \approx 14 \frac{mins}{pt}$$

Let's assume our practitioner is early in their career and wants to work no more than 10 hours a day for five days a week, with no on-call time. Then we can estimate the average daily patient volume to be:

$$Daily_{vol} = \frac{(10 \; hrs - 2 \; hrs)}{.25 \; hrs/pt} \approx 32 \frac{pts}{day}$$

This gives an annual estimated volume of:

$$Annual_{vol} = 32 \frac{pts}{day} \times 5 \frac{days}{wk} \times 50 \frac{wks}{yr} \approx 8{,}000 \frac{pts}{yr}$$

This is a large number of patients. Of course, many of these would likely be follow-up visits with regular patients. The number of patients would be lower for a neurologist or internist, since the typical neurology/internal medicine patient evaluation tends to be complex and requires more time to evaluate and treat. In contrast, a large family practice with multiple physician-expanders (physician assistants, nurse practitioners) could see a significantly larger pool of patients.

CHAPTER 28

HOW LONG IS THE AVERAGE WAIT AT A DOCTOR'S OFFICE?

For our first jaunt into more advanced topics, let's visit an issue that almost everyone can relate to—including doctors! Let's say you've obtained a referral for your migraine headaches and you've had to wait three months for an appointment. You arrive a few minutes early at the neurology clinic for an evaluation and see a full waiting room. How long will you wait to be seen?

This is a ubiquitous class of problems that deals with waiting in lines (or "queues," as they say in the UK) at hospitals, motor vehicle registration, traffic lights, buffet lines, and movie theaters, to name just a few examples. Here we must go beyond simple algebra and geometric means to make use of something called "queuing theory mathematics." For a detailed explanation of this fascinating theory, I will defer to numerous texts and online articles easily accessed via an Internet search. I personally used an old Barron's Business Review book as my reference.[26] For now, I will simply say that for the typical single-physician office, a model called "Single-Channel Waiting Line Model" works reasonably well.

The key assumptions we must use for this model are 1) the rate of patient arrivals follows something called a "Poisson probability distribution," and 2) the evaluation and treatment or "service" time follows a negative exponential distribution. These formulas are shown below, beginning with the Poisson distribution for patient-arrival rates and the negative exponential distribution for service times:

$$\text{Patient arrival rates} = P(x) = \frac{e^{-\lambda}\lambda^x}{x!}$$

$$\text{Patient service times} = P(t) = e^{-\lambda t}$$

Please consult your favorite beginning-level statistics and mathematics texts or do a simple Google search for more detail on these probability distributions.

Let's first make some key assumptions about the general rate of patient arrivals and time spent in the exam rooms. First, we'll use our prior estimate for the number of patients a family practitioner can see in a year to estimate the arrival rate; we can assume this number will also work for a neurologist's office:

$$\lambda = 30 \frac{pts}{day} \times \frac{1 \, day}{9 \, hrs} \approx 3 \frac{pts}{hr}$$

We had previously assumed that a family practitioner can see follow-up patients in about 10 minutes and new patients in 15–20 minutes. Therefore, we will estimate a general "service" time of between 10–20 minutes:

---

[26] Jae K. Shim and Joel G. Siegel, *Operations Management* (Barron's Business Review series), 1st ed. (Hauppauge, NY: Barron's Educational Series, 1999).

$$\mu = \sqrt{10 \ mins/pt \times 20 \ mins/pt} \approx 14 \ min/pt \approx 0.25 \ hrs/pt$$

Now we can use the simple single-channel queuing model to estimate a number of values that will determine how long we wait in line. First, we will calculate something known as the utilization rate, or how busy the doctor is. This is defined as the ratio of the arrival rate to the service rate:

$$Rate_{utilization} = \frac{arrival\ rate(\lambda)}{service\ rate(\mu)} = \frac{3\ \cancel{pts/hr}}{4\ \cancel{pts/hr}} \approx 75\%$$

Therefore, the probability of no one being in line before you becomes:

$$Pr_{zero\ line} = \left(1 - \frac{\lambda}{\mu}\right) \approx 25\%$$

The average number of patients in line before you (ignoring units because they cancel out) then becomes:

$$Avg_{line} = \frac{\lambda^2}{\mu(\mu - \lambda)} = \frac{3^2}{4(4 - 3)} = 2.3 \approx 2\ patients$$

The average wait time in line uses a slightly modified formula and gives us:

$$Avg_{time} = \frac{\lambda \frac{pts}{hr}}{\mu(\mu - \lambda)\frac{\cancel{pts^2}}{\cancel{hr^2}}} = \frac{3}{4(4-3)}\frac{hrs}{pt} = 0.75\ hrs$$

$$= 45\ mins$$

The math provides an exact calculation, so I did not round the numbers as I usually do. OK, so now we see that you're going to wait about 45 minutes in line to be seen, on average. How long will you wait in total to be seen and evaluated? Here the result is a little surprising and is simply the reciprocal of the difference in the service and arrival times:

$$Avg_{total\ time} = \frac{1hr}{(\mu - \lambda)\ pt} = \frac{1hr}{(4-3)\ pt} = 1hr$$

From start to finish, you'll spend at least an hour at the doctor's office.

Before we leave this case, let's visit a scenario I've seen in practice. The typical situation is that the practice manager reviews the charts and determines that the doctor's schedule is not at full capacity. What if they decide there is too much free time and squeeze a few more patients into the day to make up for no-shows and the 25% slack time? It shouldn't cause any great difficulty, should it?

This is where queuing theory begins to add value to the picture, but also illustrates where simple linear extrapolation can fail dramatically. I'll save you the calculations and illustrate the results in a table, using the assumption that the practice manager increases the number of patients per day by three patients, or 10%. This becomes 33 patients in nine hours, or about 3.7 patients per hour. The results are striking:

WAITING TIME AT THE DOCTOR

Variable	$\lambda = 3\ pts/hr$	$\lambda = 3.7\ pts/hr$	Comments
Utilization (%)	75%	93%	
Prob (zero in line)	25%	7%	
Avg number waiting	2	11	5x the number in line!
Avg wait for service	45 mins	3 hrs	Wow!

Just by adding three more patients into the day, the wait times increase non-linearly! This illustrates an important point about everyday life: many businesses and services rarely take wait times and non-linearity into account. While this may seem incredible, these are very accurate effects, which you may have experienced when seeing traffic lanes merge or check-out aisles close in the grocery store. If queuing theory is not taken into account, the result is that what appears to be an insignificant change in volume leads to adverse effects that are out of proportion to what was expected.

In real queuing modeling, we would use event-time simulation or more robust models than the simple mathematics we have worked with here. Where analytic queuing theory fails is when the arrival rate equals or exceeds the service time. In those cases, only numeric simulations can project what we have all experienced in the doctor's office and in traffic.

CHAPTER 29

HOW MANY HOSPITAL BEDS ARE THERE IN THE UNITED STATES?

H ere we can use some general information about hospitals to get a rough idea of the total number of beds in the United States. I will develop estimates for the following quantities first:

1. What is the average number of beds per hospital? Call this H_b.
2. How many hospitals are there per state? Call this H_s.
3. How many U.S. hospital beds are there overall? Call this N_b.

Having trained and worked in large hospitals, and having occasionally seen smaller ones, I will go with what I've observed from experience:

$$H_b = \sqrt{50 \times 500} \approx 160 \ beds$$

How many hospitals are there in the U.S.? Let's break this down by state and assume that there are ten. This is a small number, but at least we know it's implausible to have fewer than this. What about a maximum number? I would say more than 100, especially for states like New York or California. Could it be 1,000? It's reasonable to estimate that the number is somewhere between 10 and 1,000, or two orders of magnitude. Using the geometric mean, we will estimate an average number of hospitals per state as:

$$H_S = \sqrt{10 \times 1{,}000} \approx 100 \ hospitals \ per \ state$$

Now we can estimate the total number of hospital beds by simply multiplying our estimates together with the number of states:

$$N_b = \frac{160 \ beds}{hosp} \times \frac{100 \ hosp}{state} \times 50 \ states \approx 800K \ beds$$

So how did we do on this? The American Hospital Association published a report in January 2020 (using data from the 2018 AHA annual survey) that listed 924,107 staffed hospital beds in the U.S.[27] Our result was only about a 13% error, which is better than expected for such a simple estimate. The AHA report also showed 6,146 hospitals in the U.S., and we estimated 5,000. Again, this is a good agreement, with only about a 19% error. I hope this illustrates the utility of the Fermi approach.

---

[27] "Fast Facts on U.S. Hospitals, 2020," American Hospital Association website, updated January 2020, https://www.aha.org/statistics/fast-facts-us-hospitals.

CHAPTER 30

HOW LONG WOULD IT TAKE TO INOCULATE THE U.S. POPULATION?

A deadly flu is coming, and the Centers for Disease Control and Prevention (CDC) needs to inoculate the U.S. population quickly. How long would such an effort require? To answer this, we need to make some assumptions and estimate or look up several pieces of information. Let's assume we have a plentiful supply of influenza vaccine and that inoculation clinics can be set up quickly at local hospitals across the country. Now we need to estimate some additional pieces of information:

1. What is the number of hospitals in the U.S.? Call this the inoculation site number S_n.

2. How long does it take to inoculate someone? Call this inoculation time I_t.

3. How many inoculating staff members are at each site to give injections? Call this quantity N_s.

Hopefully we have done enough estimations now that you are becoming familiar with the approach. As we showed in Chapter 29, we can estimate the average number of hospitals per state (between 10–1,000) and multiply by 50 to get an estimate of the total within a factor of 10:

$$S_n \approx 50 \times \sqrt{10 \times 1{,}000} \approx 50 \; \sout{states} \times 100 \; hospitals \; \sout{per\;state}$$

$$\approx 5{,}000 \; sites$$

This result was quite close to the actual number of U.S. hospitals as of January 2020: 6,146.

Now let's estimate the time it takes to inoculate someone. From my own experience, this usually takes from about 2–10 minutes, with the majority of the time taken up by filling out forms, asking about allergies, etc. This results in an overall estimate of:

$$I_t = \sqrt{2 \times 10} \approx 4 \; minutes \; per \; inoculation$$

Finally, let's assume that each hospital inoculation site has at least one staff member to perform inoculations, up to a maximum of 10 staff members for the largest facilities (perhaps RNs and MDs assist when available). This gives us an average per site of:

$$N_s = \sqrt{1 \times 10} \approx 3 \; inoculation \; staff \; per \; site$$

With these quantities, we can now estimate how long it could potentially take to inoculate the entire U.S. population of approximately 300 million:

$$T_{Inoculate} = 300 \times 10^6 \, \frac{\cancel{inoc}}{\cancel{US\,pop}} \times \frac{4 \, \cancel{mins}}{\cancel{inoc}} \times \frac{\cancel{US\,pop}}{5{,}000 \, \cancel{sites}} \times \frac{1 \, \cancel{site}}{3}$$

$$\times \frac{1 \, \cancel{hr}}{60 \, \cancel{mins}} \times \frac{1 \, day}{8 \, \cancel{hrs}} \approx 170 \, days$$

Wow, that's almost six months! Could this be decreased? It probably could by making this an all-hands-on-deck national effort and recruiting clinics, schools, and churches to the effort. Let's assume this is possible and the number of sites increases to three times the number of hospitals, or 15,000 total inoculation sites. Then the result would become:

$$T_{Inoculate} = 300 \times 10^6 \, \frac{\cancel{inoc}}{\cancel{US\,pop}} \times \frac{4 \, \cancel{min}}{\cancel{inoc}} \times \frac{\cancel{US\,pop}}{15{,}000 \, \cancel{sites}} \times \frac{1 \, \cancel{site}}{3}$$

$$\times \frac{1 \, \cancel{hr}}{60 \, \cancel{min}} \times \frac{1 \, day}{8 \, \cancel{hr}} \approx 57 \, days$$

Much better, but you can see that unless significant resources are mobilized, it still takes a significant amount of time to achieve such broad coverage. Is this a realistic estimate? Think of the natural disasters in the U.S. and abroad such as hurricanes, earthquakes, and tsunamis. From these examples we can see that while initial responders can usually arrive quickly on site and work swiftly, it can easily take months to fully respond to all the health and security needs of a disaster on a large scale. Also, U.S. inoculation efforts during the annual flu season take several months to vaccinate only a portion of the at-risk population. Given this information, these numbers appear quite realistic.

CHAPTER 31

BANGING HEADS TOGETHER

Football is one of the great American sports. In recent years, though, the medical sequelae of sports-related head injuries have come into the open, resulting in guidelines and oversight for evaluating players after head injuries on the field. The good news is that physicians, coaches, and players are now recognizing the "cognitive cost" of these injuries and are implementing treatment protocols and guidelines.

Let's examine head injuries from a purely clinical, scientific perspective. What is the force of impact for a typical football head injury? How would we estimate this? We will use classical Newtonian physics for inelastic impacts to evaluate this for two scenarios: first, for a player hitting an immovable wall with his head, and second, for two players colliding with each other.

Here's a quick refresher on Newton's second law of motion: The acceleration of an object is proportional to the net force acting on the object and its mass. This is often written in scalar form as:

$$Force = mass \times acceleration$$

For our purposes, we can use this law to calculate the force of an impact on an object and its acceleration (or change in velocity). We can make a simple model of the mass (M) of our object (football player), moving with a velocity (V) toward a wall:

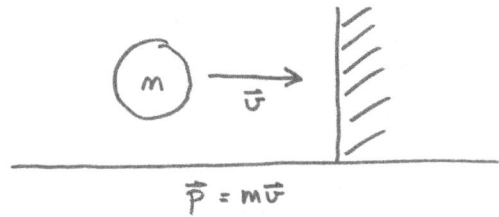

Figure 8. Football player momentum

Next, we write Newton's second law as an equation and use elementary calculus to develop the expression for a change in momentum:

$$Force = mass \times acceleration = F = ma = m\frac{dv}{dt}$$

Here the force of impact (F) of the object on the wall equals the mass times the change in velocity (the acceleration). For our purposes, we will approximate the derivative as follows:

$$F = m\frac{dv}{dt} \rightarrow F \approx m\frac{\Delta v}{\Delta t}$$

Assuming the final velocity is zero (the player is completely stopped by the collision), we can now solve for the impact force after estimating the football player's velocity and impact time. A full sprint is certainly beyond a fast walk of five miles an hour, and less than a factor of 10x higher at 30 miles an hour. Thus, we can approximate this as:

$$v = \sqrt{5 \text{ mph} \times 30 \text{ mph}} \approx 12 \text{ mph}$$

The time for a collision is probably more than a tenth of a second and likely less than one half of a second:

$$t = \sqrt{0.1 \text{ sec} \times 0.5 \text{ sec}} \approx .2 \text{ sec or } 200 \text{ msec}$$

If our player is a typical linebacker at 220 pounds, we can now convert our estimated values into the MKS system to proceed with our calculation of the impact force:

$$m_{player} = 220 \text{ lbs} \times \frac{1 \text{ kg}}{2 \text{ lbs}} \approx 110 \text{ kg}$$

Using a similar approach for the player's velocity, we have:

$$v_{player} = 12 \frac{\cancel{miles}}{\cancel{hr}} \times \frac{1 \cancel{hr}}{3,600 \text{ secs}} \times \frac{1,600 \text{ m}}{\cancel{mile}} \approx 5 \frac{m}{sec}$$

Finally, we can complete our calculation to determine the impact force as follows:

$$F_{impact} = m \frac{\Delta v}{\Delta t} = 110 \text{ kg} \times \frac{\left(0 - 5 \frac{m}{sec}\right)}{0.2 \text{ sec}} \approx -2800 \frac{kg \, m}{s^2}$$

$$\approx -2.8 \text{ kN}$$

Here I used the conversion that a kg-m/s² is one Newton (N) of force. The minus sign indicates that the direction of the force is opposite that of the player's motion. So how big is this force? We can answer that by finding and comparing the force magnitude to a common measure of acceleration forces for cars, impacts, and missile flights: the gravitational force equivalent, or G-force. This is the force an individual experiences at the earth's surface due to gravity:

$$F_{gravity} = mg = 110 \; kg \times 10 \frac{m}{s^2} \approx 1100 \frac{kg \; m}{s^2} \approx 1.1 \; kN$$

Dividing the force of head impact by the force of gravity indicates that our player is experiencing almost three Gs!

Fortunately, most individuals don't typically run head-first into stationary walls at full speed, so you can say that this example is a little artificial. What about active players colliding on a field during a game? Here the situation is slightly different, as shown in the figure below. To illustrate the situation, let's assume we have two players running toward each other for a completely inelastic collision (a tackle). Each player has the same velocity (V) as in the previous example, but in opposite directions, and each player has the same mass (M) as in the previous example. We can use our same equation of motion to estimate the force of impact:

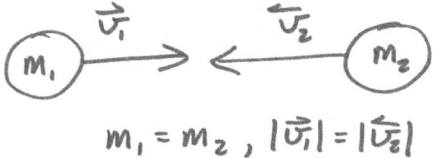

Figure 9. Two player system momentum

The final velocity of the system (both objects) will be zero, because the players are stationary after colliding. The result is that the force of impact is the force of the system over the collision time, which becomes:

$$F_{impact} = m_1 \times \frac{(v_{1f} - v_{1i})}{t_1} + m_2 \times \frac{(v_{2f} - v_{2i})}{t_2}$$

$$F_{impact} = 110 \; kg \times \frac{\left(0 - 5 \frac{m}{sec}\right)}{0.2 \; sec} + 110 \; kg \times \frac{\left(0 - 5 \frac{m}{sec}\right)}{0.2 \; sec}$$

$$\approx -5{,}600 \frac{kg\, m}{s^2}$$

Here again, the minus sign indicates that the force is in opposition to the direction of motion. As you can see here, the impact force is almost 6 Gs, which is 2x the first scenario!

To understand the significance of that number, consider these examples: A Bugatti Veyron, accelerating from 0 to 100 km/h in 2.4 seconds, has a g-force of merely 1.18, and Maxx Force at Six Flags Great America, the fastest roller coaster in 2019, has an estimated g-force of 4.5. In 2018, a Russian Soyuz rocket malfunctioned shortly after launch; its uncontrolled free fall subjected the two crew members to 7 Gs.[28] Cornering in a Formula One car can sometimes put drivers through nearly 8 Gs of lateral acceleration,[29] and highly skilled jet pilots can get to 10 Gs in .03 seconds.[30] G-forces this high can cause loss of consciousness.

The bottom line here is that these are significant forces, and players experiencing these types of head injuries multiple times per game over the span of a career are at high risk for repetitive brain injury, which can result in behavioral changes, poor impulse control, cognition changes, and even dementia.

---

[28] Denise Chow, "Soyuz Astronauts' Emergency Descent Was a Harrowing, High-G Ordeal," NBC News website, October 11, 2018, https://www.nbcnews.com/mach/science/astronauts-emergency-descent-was-harrowing-high-g-ordeal-ncna919246.

[29] James Gilboy, "Modern Formula 1 Car's Braking Forces So Brutal It'll Extract Tears from Your Eyes, Report Says," The Drive, May 24, 2019, https://www.thedrive.com/accelerator/28189/modern-formula-1-cars-braking-forces-so-brutal-itll-extract-tears-from-your-eyes-report-says.

[30] "Withstanding the High-G," Red Bull Air Race website, February 28, 2019, https://airrace.redbull.com/en/news/withstanding-high-g.

CHAPTER 32

THE SUM OF ALL IMPACTS

Now that we've estimated the g-force in a single football head impact (HI), let's figure out how many head impacts could be expected over a football career spanning teenage years to early 40s. We can use the same techniques to make reasonable estimates of numbers sustained over a professional career. We will need the total exposure time (years of play, games per year, plays per game) and the risk of HI per play to calculate this. We must recognize that some players, such as centers, are probably at greatest risk of HI, whereas quarterbacks and kickers are at much lower risk. Knowing this, let's estimate the numbers for a hypothetical center and quarterback (QB) by approximating the parameters of a typical game.

American football consists of four quarters of 15 minutes each for one hour of continuous play. The offense and defense of each team lines up for a play, which can last for a few seconds (false start) up to perhaps one to two minutes (run to the end zone). Using these approximations, we can estimate the mean duration of a play as:

$$D_{play} = \sqrt{15 \ secs \times 120 \ secs} \approx 40 \ secs = .7 \ min \approx 1 \ min$$

Given this, we can estimate the number of potential plays per game as:

$$N_{play} = \frac{60 \ \cancel{mins}/game}{1 \ \cancel{min}/play} \approx 60 \ plays/game$$

Thus, for a given center or QB, there are about 30 opportunities (one half of the total) to experience a head injury in a given game.

Now we need to estimate the potential HI events per play for each of our two positions. Centers are the central point of contact and "wall" that protects the QB from the defense, so it's reasonable to assume that the center is likely to experience a potential HI with each play (1 play = 1 HI). In contrast, from my personal experience viewing college and pro games, the QB is sacked or stopped while running the ball maybe every ten plays or so. Thus, I'll assume that the QB risk is only 10% (10 plays = 1 HI) during a tackle.

We still need two more pieces of information before we can estimate the career HI numbers: the number of games per year and the number of years of play. Let's look at the latter quantity first. I'll assume the majority of players begin serious play in their early teenage years, so let's choose 12 as the beginning age. I think we can assume four years of college football (ages 18–22) and an average professional career duration of three to six years. Based on this information, the maximum age is unlikely to extend much past late 20s or possibly 30s. I'll estimate the cumulative career by simply adding up these separate time periods:

$$D_{career} = 6 \ years \ (12 \ to \ 18 \ years) + 4 \ years \ (college) \\ + 6 \ years \ (pro) = 16 \ years$$

For players, football season typically begins with spring training, where practice games provide almost as many opportunities for HI as regular-game play. The season typically ends before Christmas, but in some cases goes into January with bowl games. Let's assume the season extends from April through January, or nine months. During the season, each team typically plays one game per week; we'll assume this frequency extends throughout the season, but our results should still be accurate within our standard 10x error assumption.

Now we can proceed to calculate estimates for our center and QB, using the values we have calculated so far:

$$Games_{year} = 1 \, \frac{game}{\cancel{week}} \times \frac{4.3 \, \cancel{weeks}}{\cancel{month}} \times \frac{9 \, \cancel{months}}{year} \approx 39 \, \frac{games}{year}$$

We can convert this to lifetime career exposure as follows:

$$Plays_{career} = 39 \, \frac{\cancel{games}}{\cancel{year}} \times \frac{16 \, \cancel{years}}{career} \times \frac{30 \, plays}{\cancel{game}}$$
$$\approx 20{,}000 \, \frac{plays}{career}$$

Returning to our team position estimates of the risk of HI per play, we can now estimate the career HI potential as:

$$nHI_{Center} = 20{,}000 \, \frac{\cancel{plays}}{career} \times \frac{1 \, HI}{\cancel{play}} \approx 20{,}000 \, \frac{HI}{career}$$
$$\approx 10^4 \, \frac{HI}{career}$$

In contrast, players with lower contact rates per play have a significantly lower number of potential HIs over a career:

$$nHI_{QB} = 20{,}000 \, \frac{\cancel{plays}}{career} \times \frac{0.1 \, HI}{\cancel{play}} \approx 2{,}000 \, \frac{HI}{career} \approx 10^3 \, \frac{HI}{career}$$

Although the lower number is a tenth of the higher one, 2,000 head injuries is still an enormous number—even 100 HIs in a lifetime is not a good thing. Of course, the actual risks could be up to 10x lower than my rough estimates here, (1,000 and 100 career head injuries), but even those lower numbers suggest that some of these young men have the potential to incur significant neurologic injury and concurrent disability over a lifetime of active play.

CHAPTER 33

PLAUSIBLE CLINICAL TRIAL ENROLLMENT

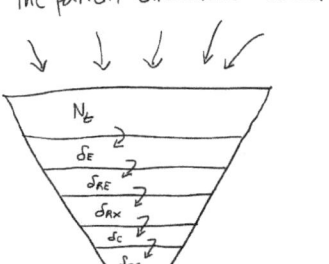

I magine you are planning a clinical trial. This could be part of an academic study of a new disease treatment or pharmaceutical evaluation for a drug, or it may be performed by governments or other organizations. All forms of clinical trials involving patients will require some method to recruit participants, and effective planning is critical for executing trials successfully. In this chapter I will assume a neurology study evaluating patients with refractory epilepsy, where the term "refractory" is defined as an individual who has not achieved seizure control after a trial of two to three generally effective anti-epileptic drugs.[31]

---

[31] Patrick Kwan and Martin J. Brodie, "Early Identification of Refractory Epilepsy," *New England Journal of Medicine* 342 (2000): 314–319, DOI:

Your first step is to estimate the plausible number of patients who might be candidates for the trial. Patients not only need to fit the potential enrollment criteria, they must also have access to a participating site, fit the criteria for the treatment, be willing to enter the study, successfully pass screening, and finally enroll in the trial. Let's assume you are planning to conduct the study at 15 sites in the U.S. at major population centers over one year. How might you estimate the plausible enrollment you could achieve?

The approach I present here is based on thinking of patient selection as a funnel: at the broad end you have the target population, and at the narrow end is the number of patients that make it through the funnel to your trial in the specified time. I'll borrow an approach developed by Frank Drake in 1961 and used in astrophysics to estimate the number of potential intelligent civilizations in the galaxy or universe, commonly referred to as the Drake equation. I will adopt much of the nomenclature and adapt it to the purpose of clinical trials, as shown below:

$$N_S = N_T f_E f_{RE} f_{Rx} f_C f_{SS} f_E$$

Where: N_S = total patients entering the study
N_T = total population in study catchment area
f_E = fraction of N_T who have epilepsy
f_{RE} = fraction of f_E who have refractory epilepsy
f_{Rx} = fraction of f_{RE} who qualify for the study
f_C = fraction of f_{Rx} who consent to be in study
f_{SS} = patients who successfully pass screening
f_E = patients who actually enroll

Let's begin with the total population covered by our planned sites. Let's say that each of our 15 sites is located in major population areas. In the U.S., major cities have a population of 1–10 million, so we'll estimate an average population catchment per site as:

10.1056/NEJM200002033420503.

$$N_{Site} = \sqrt{(1 \times 10^6) \times (10 \times 10^6)} \approx 3 \times 10^6 \text{ population per site}$$

Given 15 sites, the total population is simply 15 times our average population per site, or 45 x 10⁶ individuals. The general U.S. occurrence of epilepsy is roughly 1%, so our first step in the funnel becomes:

$$f_E = 10^{-2} \times (45 \times 10^6) \approx 45 \times 10^4 \text{ individuals}$$

Similarly, we need to estimate the number of people who have refractory epilepsy. Most patients can be managed on monotherapy (a single drug), so I would expect that the proportion of refractory patients is less than 50%. Here I deviate from my typical approach and go to a neurology board review reference to remind myself how "refractory" is defined.[32] Per Rolak, about 47% of patients are seizure-free with one medication, an additional 13% are seizure-free after two medications, and about 1–3% experience additional benefit with more medications. Using these figures, we can estimate the proportion of refractory patients as:

$$p_R = 1 - (.47 + .13 + .02) \approx .4 \text{ or } 40\%$$

Note that this result is not far from my original guess of less than 50%. Now we can estimate our refractory patient population as:

$$f_R = .4 \times (45 \times 10^4) \approx 2 \times 10^5 \text{ individuals}$$

Let's proceed with the next steps in our filter. The proportion of patients who could qualify for treatment is unlikely to be more than 50% (due to other medical problems, for instance) and not less than 1%, so we will use the geometric mean to estimate our proportion as:

$$p_Q = \sqrt{1\% \times 50\%} \approx 7\%$$

Our qualifying patient population becomes:

$$f_R = (7 \times 10^{-2}) \times (2 \times 10^5) \approx 1 \times 10^4 \text{ individuals}$$

Similarly, we will estimate the proportion of patients willing to

---

[32] Loren A. Rolak, *Neurology Secrets*, 5th ed. (Philadelphia: Mosby, Inc., 2010).

consent to being in the study as approximately 7%, reaching a consenting study population of:

$$f_C = (7 \times 10^{-2}) \times (1 \times 10^4) \approx 7 \times 10^2 \; individuals$$

Finally, we must estimate the proportion of patients who successfully pass the screening period and enroll in the trial. Drawing on my experience running multiple trials, I know the screen failure rate is typically about two thirds, or 67%. Therefore, the filter would correspondingly be 33%. Note that here I did not estimate the value from a range but from my experience.

One could also use the range method (with the geometric mean) to obtain a plausible value. The result is that we now have a potential population of:

$$f_C = (33 \times 10^{-2}) \times (7 \times 10^2) \approx 2 \times 10^2 \; individuals$$

Having started out with 15 sites over one year at large metropolitan areas in the U.S., we have filtered our plausible study population down to about 200 patients. If we assume that individuals learn about the study and become enrolled over the course of the year, this becomes a potential enrollment rate of:

$$R_{enroll} = \frac{2 \times 10^2 \; patients}{15 \; sites \times 12 \; months} \approx 1 \; patient \; per \; site \; per \; month$$

After all this work, the enrollment rate may not sound impressive. However, based on my own time writing, evaluating, and overseeing clinical trials, this is a very typical result. In truth, the enrollment rates are often far lower than this.

The approach I used to estimate enrollment rates is different from the approach typically used in pharma and academia, which estimates enrollment rates by asking individual sites how many patients could qualify. These results are often highly optimistic, and many trials fall far short of the promised patient volumes and recruitment times. In my experience, real clinical trial enrollment looks much more like the numbers I have generated here. The value of breaking down the enrollment rate using this method is that it makes all the assumptions and steps clear. Thus, one can clearly estimate early in planning whether enrollment may be a problem. This is particularly important when evaluating treatments for rare diseases; finding and recruiting individuals for your study may be difficult.

CHAPTER 34

WHAT IS THE AVERAGE U.S. INDIVIDUAL INCOME?

$$\$_{Avg} = \frac{\sum_{i=1}^{N_{America}} \$_i}{N_{America}}$$

To begin tackling problems related to healthcare costs of specific diseases, absenteeism, and so on, we need to have a basis for comparison. For many problems related to the cost of caring for illness, we use the average U.S. income. I don't know the official average wage, nor do we need it here. We can easily estimate this with the techniques we have been using and check it against the yearly income of someone making minimum wage; the average income should be greater than that, but probably not more than by a factor of two or three.

Let's start by estimating the upper and lower boundaries of the U.S. individual income range. Having filled out a number of my own tax forms, I remember that the upper end is around $250,000. I also recall this being about the number that so many political pundits refer to as the "one percenters." What's the bottom end? Zero is the absolute minimum, but that essentially ignores the population that comprises the working poor. Let's assume the lower boundary of income in the U.S. is around $10,000, which is close to the 2020 poverty level of $12,760 for an individual.[33] This gives us an annual income estimate of:

$$Income_{avg} = \sqrt{\$10{,}000 \times \$250{,}000} \approx \$50{,}000$$

This seems like a high number to me. How does it compare to the hourly minimum wage? Recently, some cities have passed laws requiring a $15-an-hour federal minimum wage. How does our average compare with this? If we use a standard person-year of 2,080 hours (40 hours/week x 52 weeks/year), the new minimum wage implies a minimum annual per-capita salary of about $31,000. My figure is a factor of 1.6, or 60%, greater than this minimum, so this is probably a reasonable estimate given the simple method we are using.

Let's dig a little more and look up current income figures. One article on historical and current income states that the *median* income per capita in 2018 was $33,706, while the *mean* income per capita was $50,431.[34] Wow, we came out very close! This information is also available via the census website.

On a side note, "income per capita" is the average for the population 15 years and older. Note the difference between the $50,431 mean and the $33,706 median (50th percentile, or middle) value. When the mean is much higher than the median, this indicates a right-skewed distribution. This distortion of the distribution of income and wealth is common throughout the world. Italian economist Vilfredo Pareto first

---

[33] "HHS Poverty Guidelines for 2020," U.S. Department of Health & Human Services website, January 8, 2020, https://aspe.hhs.gov/poverty-guidelines.

[34] Kimberly Amadeo, "Average Income in the USA by Family and Household," The Balance, updated February 17, 2020, https://www.thebalance.com/what-is-average-income-in-usa-family-household-history-3306189.

analyzed the phenomenon in the early 1900s; these highly skewed results are now known as Pareto distributions.

CHAPTER 35

WHAT IS THE LEVERAGE OF THE AVERAGE U.S. FAMILY?

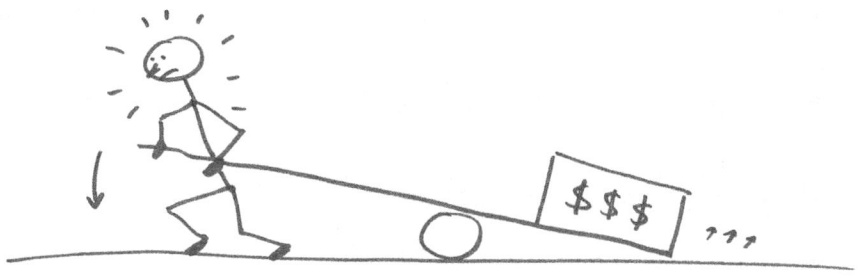

W hat is leverage? A common way to define this uses the "assets to equity" (A/E) or "debt to income" (D/I) ratio. How do we estimate this for an average U.S. household?

First, we begin by using the accounting definition: liability + equity = assets. If you're not familiar with these terms, I recommend *Finance*, an elementary financial reference.[35] Our goal now is to use our background knowledge to estimate the quantities that go into these categories.

Let's begin with *liability* (here I am using debt as the only form).

---

[35] Ehsan Nikbakht and Angelico Groppelli, *Finance* (Barron's Business Review series), 7th ed. (Hauppauge, NY: Barron's Educational Series, 2018).

The typical American family (and many international families) might owe money on a home, two cars, credit cards, and perhaps student loans. *Assets* are things that have value and could be converted into cash: the home, the two cars, any bank assets (cash, stocks, and bonds), and an income. *Equity* is simply the difference between liability and assets.

The approach we used to estimate the average U.S. household income salary will also work to estimate the other quantities, based upon our background knowledge. First, we will estimate the average home value, knowing that homes in the San Francisco Bay Area are in the millions, and homes in the central U.S. are typically a few hundred thousand at most:

$$Home_{avg} = \sqrt{\$100{,}000 \times \$1{,}000{,}000} \approx \$300{,}000$$

Next, the typical non-luxury car cost in 2020 ranged approximately from $10,000–$50,000. Therefore, our geometric mean estimate would be:

$$Car_{avg} = \sqrt{\$10{,}000 \times \$50{,}000} \approx \$22{,}000$$

Many individuals and families have very little savings and retirement (note that I am lumping retirement savings and individual savings accounts into one group). Let's use two extremes for our estimates: savings between $100 and $100,000.

$$Savings_{avg} = \sqrt{\$100 \times \$1{,}000{,}000} \approx \$10{,}000$$

Now we go to the debt side of the balance sheet. My impression from the financial news is that many families have high debt loads. Rather than estimate the actual dollar value of debt directly, I'll estimate the proportion that is paid off, which is generally small.

$$Debt\ retirement\ (\%) = \sqrt{1\% \times 100\%} \approx 10\%$$

We can use this figure to estimate how much debt remains on the value of the cars and home from above:

$$Home\ debt = \$300K \times (1 - 10\%) \approx \$270K$$

$$Car\ debt = \$20K \times (1 - 10\%) \approx \$18K$$

To calculate the debt levels for credit cards and student loans, I'll estimate the average from the following plausible ranges:

$$Credit\ cards = \sqrt{\$100 \times \$30{,}000} \approx \$2{,}000$$

$$Student\ loans = \sqrt{\$1{,}000 \times \$100{,}000} \approx \$10{,}000$$

Now we are in a position to populate a balance sheet for an American family with two cars and two $50K/year incomes as follows, with the equity being the difference between assets and liabilities (cumulative debts):

AVERAGE FAMILY BALANCE SHEET

Assets	Debts
Savings $10k	Credit card $2k
Salary $100k	Car $20k
Home $300k	Home $270k
Car $22k	Student loans $10k
	Equity
	$130k

This actually looks pretty good and may be an overly optimistic balance sheet. So what's the financial leverage? Here we will use consumer leverage ratio = debt/income, which in our case becomes:

$$CLR = \frac{debts}{salary} = \frac{\$300K}{\$100K} \approx 3x$$

This is actually quite high relative to published data from the U.S. Federal Reserve and Bureau of Economic Analysis. In 2016, the figure was 1.04x; in 2007, it was at an all-time high of 1.29x. However, our estimate is only off by about a factor of three, which is quite good given the rough assumptions we are using.

CHAPTER 36

WHAT IS THE ANNUAL COST OF U.S. HEALTHCARE?

What does the average American spend on healthcare annually? We could just Google this, but it's fun to show that we can estimate this sum quite well with just a few simple pieces of information. I'm going to go into some detail here to reflect costs, but I think the process will be worth it.

Let's start by using our previous estimate for the frequency of outpatient visits when we estimated the total number of U.S. doctors. There we took the geometric mean of visits between 1–12 times per year:

$$Visit_{freq} = \sqrt{1 \times 12} = 3.5 \approx 4 \; visits \; per \; year$$

What does the average visit cost? Using my memory of the high and low values for simple follow-up visits and very complicated new visits, I will estimate the geometric mean:

$$Visit_{cost} = \sqrt{\$100 \times \$750} = \$274 \approx \$300 \; per \; visit$$

Note that I am assuming the "cost" to the system or patient is equivalent to the billing charge, which is not exactly true due to discounts, insurance coverage, etc. However, it is not too far off. Making this assumption gives a total annual estimated cost for outpatient care:

$$Annual \; cost_{op} = \frac{\$300}{\cancel{visit}} \times \frac{4 \; \cancel{visits}}{year} \approx \$1,200 \; per \; year$$

Now let's turn to inpatient costs. Here we need to estimate a few more terms, including the average number of hospital stays per year per patient, average length of stay, and average cost per stay. Thankfully, hospitalizations tend to be relatively infrequent for the average person. Let's assume the typical hospital patient is admitted between one and four times per year:

$$N_{hosp} = \sqrt{1 \times 4} = 2 \; hospitalizations \; per \; year$$

While a few patients remain hospitalized for extended periods of time, many go home quickly. We will assume that the typical length of stay (LOS) varies between 1–30 days for longer admissions. We then use this range between the shortest and longest LOS to estimate the geometric mean:

$$LOS = \sqrt{1 \times 30} \approx 6 \; days \; per \; hospitalization$$

Next we will estimate the cost per stay. Hospital costs tend to be opaque, but we know that insurance coverages list payments ranging from $1,000 a day to much higher for complicated surgeries and stays in ICUs. Let's assume the costs range between $1,000–$10,000 per day:

$$Cost = \sqrt{\$1,000 \times \$10,000} \approx \$3,000 \; per \; day$$

Finally, we can develop an estimate for annual hospital cost per individual:

$$Annual\ cost_{hosp} = \frac{2\ \cancel{hosp}}{year} \times \frac{6\ \cancel{days}}{\cancel{hosp}} \times \frac{\$3{,}000}{\cancel{day}}$$

$$\approx \$36{,}000\ per\ year$$

Most of the population is not hospitalized every year. How many are? I will make a reasonable assumption that it is somewhere between 1–100%, so let's use the geometric mean as our estimate:

$$Proportion_{hosp} = \sqrt{1\% \times 100\%} \approx 10\%$$

Now we can use our estimates to derive the annual cost per individual. We put everything together as follows:

$$Annual\ cost_{per\ person} = \frac{\$1{,}200}{\cancel{year}} + \frac{\$36{,}000}{\cancel{year}} \times (10\%) = \$4{,}800$$

$$\approx \$5{,}000\ per\ year$$

So our estimate is about $5K per person in 2016. How did we do? According to one estimate, the typical U.S. worker in 2017 spent between $4,500 and $8,300 on healthcare, while the per-capita cost was higher, at $10,200.[36] We are certainly very close to the range for the workers and are only 2x different for the per-capita result quoted above, so this is quite a good estimate.

Finally, if we compare our estimated per-person expenditure to the average health premiums paid in 2020 (ValuePenguin), we can see these range between $3,432–$13,476, for a geometric mean of $6,387. Our result is less than 1x different from the 2020 premiums!

---

[36] Tanza Loudenback, "The Average Cost of Healthcare in 21 Different Countries," Business Insider website, March 7, 2019, https://www.businessinsider.com/cost-of-healthcare-countries-ranked-2019-3.

CHAPTER 37

WHAT IS MY HEALTH INSURANCE PREMIUM?

For many of us, enrollment in our company's health plan or government-sponsored insurance program is an annual ritual. Insurance costs and premiums are a common topic of public debate and news in the healthcare arena, but the simple mathematics underlying insurance is almost never discussed. This is unfortunate, because a basic understanding of the supporting principles would quickly terminate many debates about healthcare affordability.

Let me illustrate. Imagine we have formed an insurance company, ABC Health, that provides one-year policies to customers. We have two policies for two customer populations based on age, and each policy guarantees payments to providers for claims incurred by covered customers over the period of the contract. These are also known as "sickness funds." The essential concept of insurance is that health events occur randomly and independently among members of a large insured population. In other words, if Mr. Jones develops gout during a year and requires $X of treatment, his situation does not affect Ms. Lane's probability of making a claim for her sinus infection requiring $Y of treatment. Assuming that the population covered is large (perhaps in the millions) and heterogeneous, such that claims occur randomly across the population and do not affect other individuals making claims, then premiums paid in by the under-utilizing members will cover the costs of the over-utilizing members. This will become clear with the following example.

Using our two policies as defined above, as well as our estimated annual per-capita healthcare expenditure estimate from the previous chapter, we have:

- Policy 1 (P1)
  - N = 1,000
  - Age range = 20–40
  - Average one-year health expenses = $5,000

- Policy 2 (P2)
  - N = 250
  - Age range = 40–70
  - Average one-year health expense = $10,000 (assuming premiums for this age group are about 2x the 20–40 group)

For us to remain solvent, we must charge overall premiums that ensure the health expenses of our covered population's administrative and overhead while still making a profit. Since this is a one-year contract, we will ignore the time value of money and interest compounding that would be necessary for longer-term contracts such as life insurance. We will also ignore deductible levels, which will lower premiums as the individual or family takes on more of the financial risk by paying a portion of the cost up front.

Finally, it should be noted that the system works as long as claim costs are "well-behaved," or not extreme. Unfortunately, extraordinary claims can occur, and even a small proportion of covered individuals experiencing severe health events requiring very high treatment costs can render an insurance fund insolvent. Therefore, we will add in a buffer to our premium to allow for this. This is essentially "insurance" for the fund and can also be achieved by "re-insuring" our funds with another level insurer.

Now we can estimate what our premium should be by writing down an equation with the variables we have discussed above:

$$Premium = Q \times (1 + (F_s + C_a + P))$$

Here Q = expected annual health cost (claim) per policyholder, F_s is the safety factor, C_a is the administrative costs, and P is the profit. First, we need the expected annual health cost for our covered populations. If each individual (i) in a population of N individuals makes an annual cumulative claim (y), we can estimate the expected cost of the population as the arithmetic average:

$$Q = average\ annual\ claim = \sum_{i=1}^{N} \frac{y_i}{N}$$

What should our safety factor be? Here I will use the Fermi approach and take the geometric mean of estimates, assuming this should at least be more than 1% but likely less than 50% of the expected annual cost (Q):

$$F_s = \sqrt{1\% \times 50\%} \approx 7\%$$

From my background, I know that administrative and overhead costs are typically greater than 5% and less than 25%, so I will use these to estimate our administrative costs:

$$C_a = \sqrt{5\% \times 25\%} \approx 11\%$$

From reading newspapers and magazines, I've learned that insurance firms have relatively modest profit ranges between 1–10%:

$$C_a = \sqrt{1\% \times 10\%} \approx 3\%$$

Now we are in a position to estimate our gross premiums for the two populations presented in the beginning of this problem. For the first policy, our gross premium is:

$$P_1 = \$5{,}000 \times (1 + .07 + .11 + .03) \approx \$6{,}000 \; per \; year$$

How does this value compare with 2020 data? According to ValuePenguin, annual premiums range between $4,356–$5,736, for a geometric mean of $5,027 for individuals 20–40 years old.[37] Our result is only off by 13%, which is quite good.

For the second policy (P_2), our gross premium estimate (using the same factors as P_1) is approximately $12,000 per year. ValuePenguin reports that individuals 40–70 years old will pay annual premiums that range between $5,736–$13,476, for a geometric mean of approximately $8,766. Our result is less than 2x greater than the official data, which is well within our 10x expected precision for the Fermi method.

Why the difference in cost for the two age groups? The difference between the premium estimates is mostly due to the underlying expected annual healthcare costs for the two groups. To keep things simple, I've listed age as the only demographic variable to consider. However, I could have included other variables such as sex, presence or absence of diabetes, and heart disease, all of which could also affect an individual's propensity to experience higher healthcare costs. This is a risk for the insurer—charging different premiums for different risk categories is called "risk-rating" or "experience-rating." What happens if a

---

[37] "Average Cost of Health Insurance Plans by Age (2020)," ValuePenguin, updated February 7, 2020, https://www.valuepenguin.com/how-age-affects-health-insurance-costs.

new regulation states that we must charge one premium rate regardless of risk? Let's illustrate this using the following table below:

GROSS PREMIUMS FOR INDIVIDUAL AND COMBINED POLICIES

Variable	Policy 1	Policy 2	Combined, Policy 3
N	1,000 (80%)	250 (20%)	1,250
Age	20–40	40–70	20–70
Q	$5,000	$10,000	$6,000
$1 + (F_s + C_a + P)$	1.21	1.21	
Premium	$6,000	$12,000	$7,200
Combined % diff	20%	-40%	

When the populations are combined, the effect is an increase in premiums for the lower-risk group (the majority), and a decrease for the higher-risk group. This should not be a surprise, as this is how insurance works. As long as the average premiums paid cover the costs of the entire covered population, all is well.

Using this information and a similar approach, we can now ask ourselves an important question that is currently in the news: What would it cost to provide health insurance for the entire U.S. population? Recall that we estimated our previously calculated per-person healthcare cost for the U.S. as approximately $5,000 per year, with 2014 data showing the actual per-capita cost was around $9,000 per year.

To be a bit more precise than usual, let's assume that current costs are approximately equal to the 2014 per-capita result. Using these figures, we can now estimate the annual per-capita cost of insurance for the U.S. population as:

$$\textit{Per capita premium} = \$9,000 \times (1 + .07 + .11 + .03)$$
$$\approx \$11,000 \textit{ per year}$$

Note that I am using the term "premium" here. If this were really a

Centers for Medicaid & Medicare Services (CMS) program, we could eliminate the profit. However, the administration and overhead costs could be higher, perhaps as much as 2x. Also, the safety factor may not be relevant, because the U.S. government would likely resort to taxation or inflation as a way to cover cost overruns.

Let's recalculate the premium, assuming that the typical inefficiencies of government-run organizations increase the administration costs more than the elimination of profit and safety factors. I will use an overhead cost of 2.5x our administration cost (11%), resulting in a total factor of about 25%. This is just a little higher than our original 21% load for the private insurance company. The net result is a total cost per capita of:

Per capita premium $= \$9{,}000 \times (1 + .25) \approx \$11{,}250 \ per \ year$

The aggregate annual cost for the entire country would then become:

$$Aggregate \ cost = \frac{\$11{,}250}{\cancel{person}} \times 300 \times 10^6 \ \cancel{persons}$$

$$\approx \$3 \times 10^{12} \ per \ year$$

Three trillion dollars! The U.S. GDP in 2017 was $19 trillion, so this is about 16% of GDP, which is very close to the approximate annual percentage of U.S. GDP spent on healthcare in 2019. Total federal outlays in 2017 totaled approximately $4T. If we assume that at least one third of the proposed $3T estimate ($1T) for healthcare is already accounted for within U.S. federal outlays (total federal government spending) for Medicaid, Medicare, and other programs, we can see that converting all health insurance to a fully covered government program at current cost levels would lead to about a 50% increase in spending: ($6T–$4T)/$4T. Even if such a single-payer system could negotiate costs down by 33%, this would still be approximately $2T annually and an increase in federal spending of $1.4T (assuming one third of the $2T is already covered in the federal budget), or an approximate 33% increase.

How do my results compare with others? I downloaded the recent Blahous paper that estimated the costs of the Medicare for All (M4A) proposal.[38] In this paper, the overall results estimate $33T over 10 years, or about $3T annually. This is very close to the results I obtained above.

The point here is simply to show that in aggregate, covering healthcare costs is a substantial economic outlay at the individual, population, or government level. Whether healthcare is financed under the form of commercial insurance or an entitlement program, the underlying costs and risks remain enormous.

---

[38] Charles Blahous, "The Costs of a National Single-Payer Healthcare System," (working paper, Mercatus Center, George Mason University, Arlington, VA, 2018).

CHAPTER 38

WHAT IS THE COST OF A MIGRAINE?

For those of you who have never had a headache (HA), count yourself lucky. As a lifelong member of the HA club, I can say that a severe headache can be life altering—even disabling. I am only speaking of tension-type headaches, the kind from which I suffer. Migraine headaches (MHAs) are of a different form and severity entirely. As a neurologist, I have treated many migraine sufferers; I hope I was able to help some of them. Knowing that headaches are common throughout the general population and having seen the severity and disruption these episodes have caused, I'm interested in estimating the cost of headaches, specifically migraines.

How might we estimate the cost? One way is to first estimate the number of migraine episodes in a population during a given time, and then estimate the duration. We can use our previous evaluations of the average U.S. household income to put a monetary cost on these events. Let's begin with the number of migraine headaches (MHAs).

We can compute an estimate for the annual number of episodes as follows:

$$N_m = P_{US} \times f_m \times P_m \times R_m$$

In this equation, P_{US} is the U.S. population, f_m is the fraction of the population that experiences migraines, P_m is the prevalence of migraines in the population, and R_m is the annual number of migraines per sufferer. Here we can use a few facts about MHAs to estimate some of the quantities: The U.S. population is 300×10^6, migraines are usually experienced in the fertile years (15 through 50s), and we see a higher prevalence of migraines in women. For our purposes, we will simply assume that the U.S. population age is linear, with 15–50 occurring across a span from 1–80, or:

$$f_m = \frac{50 - 15}{80 - 1} \approx 44\%$$

Next we need to estimate the prevalence of migraines. From my practice as a neurologist, I know that the annual prevalence ranges between about 5–20%, for men and women respectively. We will use the geometric mean to estimate the total population prevalence as:

$$P_m = \sqrt{5\% \times 20\%} \approx 10\%$$

Finally, we need to estimate the annual frequency of migraines per patient per year. I am not certain about the number, but I expect it to be at least one episode and not more than 48 (or two per month):

$$R_m = \sqrt{1 \times 48} \approx 7 \; MHAs/year$$

Now we are in a position to estimate the annual number of potential migraines in the U.S. as:

$$N_m = 300 \times 10^6 \text{ pop} \times .44 \times \frac{.1 \text{ migrainer}}{\text{pop}} \times \frac{7 \text{ HA}}{\text{year} - \text{migrainer}}$$

$$\approx \frac{8 \times 10^7 \text{ HAs}}{\text{year}}$$

This alone is a large number. So how much do these episodes cost society? Let's assume that the cost is only due to the lost productivity, because our migraine sufferer is unable to work, care for others, or perform their activities of daily living. Thus, we will estimate this cost as the product of the typical migraine duration and the average U.S. hourly wage:

$$C_m = D_m \times H_w$$

From my experience treating migraines, I know they typically last a few hours to sometimes two days or more. Here I am attempting to derive one average to describe the average duration for two clinical populations of migraine patients: those with fewer episodic events and those with more chronic headaches. Let's use four hours as the minimum (because of successful migraine treatment) and 48 hours as the maximum, assuming it ends when the patient is finally able to sleep. We then have an estimate of:

$$D_m = \sqrt{4 \text{ hrs} \times 48 \text{ hrs}} \approx 14 \text{ hrs}$$

Using our approximation for the average hourly wage, we can now estimate the indirect opportunity time cost of our event at the individual and societal levels:

$$C_m = 14 \text{ hrs} \times \frac{\$25}{\text{hr}} \approx \$350 \text{ per episode}$$

Note that this does not include the cost of any treatments, hospital admissions, etc. Our estimate is strictly limited to the indirect opportunity-time cost of the duration of time spent with the condition.

How do these results compare with the literature? I found

information on one migraine website,[39] where experts discussed the results of a recent paper on the economic burden of neurologic diseases, including migraines.[40] These results show that total annual costs for people with episodic and chronic migraine are between $2,000 and approximately $9,000 annually, or $78B, with the proportion attributable to indirect costs between 40–90%. Using our results of an average of seven events per year at $350 per event, we arrive at an estimate of $2,450 annually. If we take the geometric mean of the Gooch paper results, we can estimate the total annual costs as:

$$Total\ annual\ costs = \sqrt{\$2{,}000 \times \$9{,}000} \approx \$4{,}400$$

If we then take the geometric mean of the proportion for indirect costs, we obtain:

$$Indirect\ (\%) = \sqrt{40\% \times 90\%} \approx 60\%$$

This allows us to estimate the annual indirect costs based on the Gooch paper as $2,640, which is less than 10% different from my own estimate.

As a last step, let us estimate the societal-level duration and cost:

$$D_m = \frac{14\ hrs}{migraine} \times 80 \times 10^6\ migraines \times \frac{1\ day}{24\ hrs} \times \frac{1\ year}{365\ days}$$
$$\approx 130{,}000\ yrs$$

This is approximately 130,000 years of lost person-time annually, or about $6B in lost productivity! Using the Gooch estimate of $78B in annual costs, and the proportion attributable to indirect costs as about 60% or $47B, we can see that our result is approximately 7x different, which is still within the 10x uncertainty one could expect with a Fermi estimate.

This is a very big cost attributable only to migraines. One can easily

---

[39] Greg Bullock, "The Cost of Migraine," TheraSpecs.com, April 10, 2017, https://www.theraspecs.com/blog/what-a-new-report-reminds-us-about-the-high-cost-of-migraine/.

[40] Clifton L. Gooch, Etienne Pracht, and Amy R. Borenstein, "The Burden of Neurological Disease in the United States: A Summary Report and Call to Action," Abstract, *Annals of Neurology* 81, no. 4 (2017), doi:10.1002/ana.24897.

picture the cumulative cost when migraines are combined with other headache forms and diseases. As we can see, the economic costs of what many would consider a minor health problem can become staggering when aggregated to the entire population—and we have not even considered the psychological and personal suffering these conditions entail.

CHAPTER 39

WHAT IS THE COST OF ALZHEIMER'S CARE?

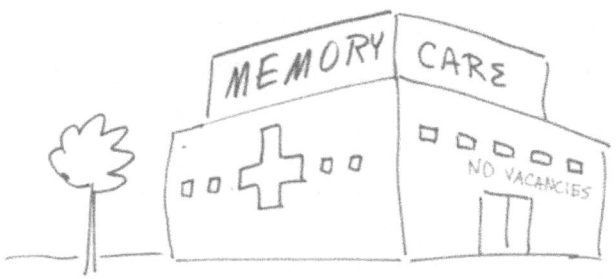

For neurologists and pharmaceutical physicians, one of the more challenging syndromes to treat is dementia, a general term for loss of memory and other mental abilities severe enough to interfere with daily life. In 2019, the history of drug development for Alzheimer's disease, the most common type of dementia, has a very poor success record. Billions of dollars and years of research have come up with only a few treatments affecting the anticholinergic pathway since the 1990s.

Currently, there are no effective treatments that can reverse or stop the disease progression, so we are left with a growing population of loved ones who develop dementia and require significant supportive care from families, the medical system, and the state. While we cannot solve these problems here, we can at least look at the problem and provide an estimate of the potential burden of this disease.

A key aspect of determining the cost of supportive care is to define what such care entails. When someone with Alzheimer's goes to a memory care facility or nursing home, they typically require nursing/tech care (getting dressed, showering, eating, activities, etc.), food, and a bed. The same care is required for someone receiving at-home care. Using our previous information and reasoning, we can estimate what these components of care typically cost for an Alzheimer's patient.

For example, we have estimated the annual U.S. individual income at approximately $50k, or $25 per hour for a 2,000-hour work year. Nurses are highly trained and typically make more than the average person (let's assume 25% more), but lower-skilled medical technicians make less than the average person (let's assume 25% less). Thus, we can estimate the average attendant hourly wage as:

$$Hourly_{avg} = \sqrt{\$25^2 \times (1.25 \times 0.75)} \approx \$24 \; per \; hour$$

Next, we estimate the food costs (note that "costs" here refer to what we as families and caregivers are asked to pay, not what the cost is to the provider). Depending on where you eat, a typical meal in the U.S. could cost as little as $10 at a fast food outlet or as much as $50 for an upper-end meal in a normal restaurant (not elegant dining). We can use these approximations to estimate the typical meal charge in an assisted-living facility:

$$Meal_{avg} = \sqrt{\$10 \times \$50} \approx \$22 \; per \; meal$$

The cost of an inexpensive hotel room can be a proxy for the cost of a room in a care facility; these can range between about $75 per night at the lower end to $250 at the higher end for a basic room:

$$Board_{avg} = \sqrt{\$75 \times \$150} \approx \$100 \; per \; day$$

How much time is required for the nursing staff? Home-care

nursing is typically used for a limited amount of time (about four hours) but is sometimes used 24 hours a day. In a care facility, some staff is available at all times, so the upper end here would be 24 hours. Now we will use this as a proxy to estimate the time for which we need trained staff each day:

$$Time_{avg} = \sqrt{4 \times 24} \approx 10 \text{ hours per day}$$

Next, we can use these estimates to pull together an average daily cost of at-home supportive care:

$$Daily_{hc} = \left(\frac{\$24}{\text{hr}} \times \frac{10 \text{ hrs}}{\text{day}}\right) \approx \$240 \text{ per day}$$

Note that I have assumed that the food and lodging is already accounted for in the home living expenses. The result is that with an average of 30 days a month, in-home care is approximately $7,000 per month, or about $80,000 per year. This is hopefully an overestimate; many home-care nurses provide care for a given Alzheimer's patient only a few days a week, for about four to six hours per day. However, this figure gives a picture of a potential upper-range cost.

What about care in a facility? Here we must revisit some assumptions. In a memory care or nursing facility, each staff member typically cares for many residents. In a hospital, this is known as a "nursing or staff ratio." Let's assume that the cost of direct staff care is shared among the number covered. From experience visiting loved ones in facilities, I've seen the number of patients to staff range between about three and ten. This is very crude, but gives us the following staffing ratio estimates:

$$PatientToStaff_{avg} = \sqrt{3 \times 10} \approx 5$$

Using this information, we can now estimate the daily cost for facility care:

$$Daily_{cc} = \left(\frac{1}{5} \times \left(\frac{\$24}{hr} \times \frac{10 \; hrs}{day}\right)\right) + \left(\frac{\$22}{meal} \times \frac{3 \; meals}{day}\right) + \frac{\$100}{day}$$

$$\approx \$200 \; per \; day$$

This works out to approximately $6,000 per month or $70,000 per year in direct costs. Not far from the amount we estimated for in-home care, and again a very substantial number when compared to the average U.S. individual income.

Finally, what would be a plausible cost for the total care of an Alzheimer's patient? Assuming that most patients with this disease die of other causes such as pneumonia, we can estimate that the range of survival extends from perhaps a few years to 20 years. This would give an estimated average survival of:

$$Time_{avg} = \sqrt{2 \times 20} \approx 7 \; years$$

So in this case, total care for the duration of the disease could range from about $500,000 to $600,000. These are substantial numbers! Extrapolated to the entire country, assuming approximately a 1% prevalence (a geometric mean of .1% and 10%) of Alzheimer's disease and requiring up to seven years of care, the totals are truly jaw-dropping: three million individuals requiring care that costs between $500–$600B ($1B = $1 x 10^9).

So how do all these estimates measure up to current costs in 2019? Two of my colleagues are paying close to $6,000 per month for a memory care facility, and another family member is paying approximately $4,000 per month for a lower-level "bronze" category care for a high-functioning loved one. These individuals are doing well and are not in full-care nursing facilities, so they may not be representative of the more costly cases. However, these numbers are less than a factor of 2x off my estimates.

I looked up some figures to see how much a care worker with the

Alzheimer's Association could make.[41] I then visited the association's website for other costs.[42] How did I do?

FERMI ESTIMATES FOR ALZHEIMER'S CARE COSTS

Quantity estimated	Fermi estimate	ALZ.org	Comparison
Care staff hourly wage	$24	$13–23	< 1x difference
Alzheimer's patients	1% or 3,000,000	5.8 million	< 2x difference
Annual cost (2019)	$75k x 3,000,000 = $225B	$290 billion	< 1x difference
Annual cost (2019)	$200 to $240 per day	$275 per day for private room in nursing home	< 1x difference

I summarized results in a table for easier viewing. In my estimations, I attempted to assess costs a bit more conservatively than I normally would; the results suggest that the Fermi estimates are quite close to the most current actuals for 2019. Good math, but not good news. These calculations show the enormous, complex problem of this disease and its effects.

---

[41] "Alzheimer's Association Hourly Pay," Glassdoor.com, last modified November 5, 2019, https://www.glassdoor.com/Hourly-Pay/Alzheimer-s-Association-Hourly-Pay-E17278.htm.
[42] "Facts and Figures," Alzheimer's Association website, accessed November 21, 2019, https://www.alz.org/alzheimers-dementia/facts-figures.

CHAPTER 40

INTRODUCTION TO BAYESIAN REASONING

$$P(H|D) = \frac{P(H) \times P(D|H)}{P(D)}$$

Until now, we have focused on simple estimation of quantities and probabilities using the Fermi approach. In Chapter 1, I briefly introduced you to Reverend Thomas Bayes. As we near the end of this book, I'd like to explain his famous proposition, Bayes' theorem. I will show that this approach can be used to diagnose patients and can be expanded to deal with multiple pieces of evidence and multiple competing hypotheses. Bayes' theorem is incredibly robust; even using very approximate probabilities, it is one of the most powerful algorithms known for converging on the truth. It has been used to find lost submarines, break the Enigma code in World War II, predict election outcomes successfully, filter spam, and perform machine-assisted diagnoses with great accuracy.[43]

I will assume that you have a basic understanding of probability. If you are unfamiliar with the concepts, please consult your favorite references or peruse one of the many excellent online tutorials on probability.

To set the stage, let me first describe two general forms of probability. One form is called "subjective probability," generally defined as the "strength of belief" about an event or concept in the larger world. For example, my subjective probability or strength of belief that the sun will rise tomorrow is 1.0 (or 100%), where my belief can range from 0.0 (certainty that something will *not* occur or isn't true) to 1.0 (certainty that something *will* occur or is true). Additionally, there is also an important concept of "relative probability," defined as the relative frequency that an event of interest occurs, such as the frequency of heads versus tails by flipping a fair coin. This latter approach is called the "frequentist" interpretation of probability and is generally the form taught in most introductory textbooks. Both subjective and relative probabilities must conform to the rules of probability.

With this prelude, let me write out the most common form of Bayes' theorem as follows, where H = hypothesis, E = evidence, and P(H|E) = the probability that H is true, given E:

$$p(H|E) = \frac{p(H)p(E|H)}{p(E)}$$

P(H) is known as "the prior," and we will interpret this as the subjective probability that H is true, given one's background or prior information. Another way of stating this is to think of the prior as an estimate of how typical our hypothesis is for explaining the event of interest. P(E|H), also known as "the likelihood," is the probability of observing E, given (or assuming) that H is true. Finally, P(E), or "the

---

[43] Sharon Bertsch McGrayne, *The Theory That Would Not Die: How Bayes' Rule Cracked the Enigma Code, Hunted Down Russian Submarines, and Emerged Triumphant from Two Centuries of Controversy* (New Haven, CT: Yale University Press, 2012).

marginal likelihood," is the marginal probability of observing the evidence. In many cases, the marginal likelihood can be ignored. The above formula can be unpacked to make it easier to use, although it may appear more complex. I will use something called the "total probability" to rewrite the marginal probability as follows:

$$P(E) = P(H)P(E|H) + P(\sim H)P(E|\sim H)$$

This essentially says that the probability of observing the evidence present is equal to the prior probability (based on your background knowledge or information) multiplied by the probability of seeing the evidence present if the hypothesis is true. This number is then added to the probability of your prior "guess" not being true [1 − P(H)] multiplied by the probability of observing the evidence present, if the hypothesis is *not* true.

Note that the first half of this equation is simply the numerator of our Bayes equation above. Also, to make the assumptions a little clearer, I like to use the approach discussed by Richard Carrier,[44] where the prior is made a little more explicit by writing it as "P(H|B)," where B is one's background knowledge or belief, or more formally as P(H|background knowledge).

We can now put this all together to rewrite the equation in a way that is a bit easier to calculate mechanically:

$$p(H|E) = \frac{p(H|B)p(E|H)}{p(H|B)p(E|H) + (1 - p(H|B))p(E|\sim H)}$$

At first, this long form of Bayes' equation may appear to have simply complicated things. However, I will work through some examples to illustrate that this formulation will allow anyone with a hand calculator (or preferably a spreadsheet) to work through simple Bayesian estimates with minimal errors. Before doing that, I will rewrite the equation in a form that may make it more intuitive. Borrowing from Carrier's book:

---

[44] Richard Carrier, *Proving History: Bayes's Theorem and the Quest for the Historical Jesus* (Amherst, NY: Prometheus Books, 2012), 50.

$$\text{The probability our explanation is true} = \frac{\begin{pmatrix}\text{How likely our explanation is true based on background knowledge}\end{pmatrix} \times \begin{pmatrix}\text{How expected the evidence is if our explanation is true}\end{pmatrix}}{(\text{Repeat the numerator}) + \begin{bmatrix}\begin{pmatrix}\text{How unlikely our explanation is true}\end{pmatrix} \times \begin{pmatrix}\text{How expected the evidence is if our explanation is not true}\end{pmatrix}\end{bmatrix}}$$

I will finish this chapter by working through a simple example inspired by James Stone's book on Bayes' Rule.[45] Let's assume the following scenario:

A patient presents in your office stating that he lost consciousness and had a seizure. This is a 27-year-old male with no prior medical history or medications. You have not examined the patient yet, but his physical appearance shows no signs of bruising or facial lacerations that might be seen with falls from generalized tonic-clonic seizures. The patient also appears to be alert and oriented to time and space, based on his presentation. Did this individual patient have a seizure?

As a neurologist, I estimate that the population prevalence of seizures in young adults referred for work-up is approximately 40% (this is *not* the overall population prevalence of epilepsy). This will be my prior belief. The patient states that he also experienced loss of consciousness (LOC). At this point, this is all I know. So now I will turn on the Bayesian crank and generate an updated probability based on this evidence. Let me walk through this specifically as follows:

a. My prior probability [p(H|background knowledge)] is 40%.

---

[45] James V Stone, *Bayes' Rule: A Tutorial Introduction to Bayesian Analysis* (Sheffield, England: Sebtel Press, 2013).

Note that I could easily be off a true frequency by as much as +/- 20%, but that is OK, as I will discuss later.

b. Based on my training and background knowledge, the probability of seeing a young patient like this with a history of LOC being called a seizure [$p(E|H)$] is ~90%.

c. In neurology practice, working up patients referred for LOC and presumed seizure is a relatively common occurrence. In my experience, the probability of seeing a patient like this presenting with this history not due to a seizure ($p(E|\sim H)$) is about 45%. Note that $p(E|\sim H) \neq 1-p(E|H)$.

d. By definition, the probability that my prior "a" is not true ($p(\sim H)$) is simply 1 minus my prior.

Now we can simply plug in numbers and crunch through the calculation:

$$p(H|E) = \frac{0.4 \times 0.9}{(0.4 \times 0.9) + (0.6 \times 0.45)}$$

The result turns out to be 0.57, or 57%. So, my probability or subjective belief that the patient had a seizure has increased from 40–57% based on this evidence. Note that in real life, no physician I know sits and mechanically calculates probabilities like this while seeing patients. Physicians are taught the "hypothetico-deductive" diagnostic process, where one ideally has multiple hypotheses (potential diagnoses) that we intuitively prune based on the evidence gained from the patient's medical history, physical examination, and other objective testing. This process has many features of Bayesian reasoning, and there is much literature discussing how the process of medical diagnosis is often a sort of intuitive Bayesian approach.[46]

---

[46] Harold C. Sox, Michael C. Higgins, and Douglas K. Owens, *Medical Decision Making* (Hoboken, NJ: Wiley-Blackwell, 2013).

Many human decisions can mimic Bayesian reasoning. However, both in medicine as well as in daily life, many of us are not "calibrated" reasoners— meaning that if we believe that something is 70% true or 15% false, the objective evidence doesn't show that our assessments were correct 70% or false 15% of the time. There is evidence that subjective probability calibration can be achieved, but it requires practice and feedback.[47]

In summary, this is an extremely brief, mechanical introduction to the wonderful world of Reverend Bayes. Here I illustrated how this could be used to make an initial assessment of how likely an initial diagnosis may be based on one's background information, a visual impression, and medical history. This same approach can also be used to assess simple items of interest in the news or everyday life.

---

[47] Luis Dias, Alec Morton, John Quigley, eds., *Elicitation: The Science and Art of Structuring Judgement*, vol. 261, International Series in Operations Research & Management Science (New York: Springer, 2017).

CHAPTER 41

BAYES MEETS SHAKESPEARE'S MONKEYS: MONTE CARLO BAYESIAN ANALYSIS

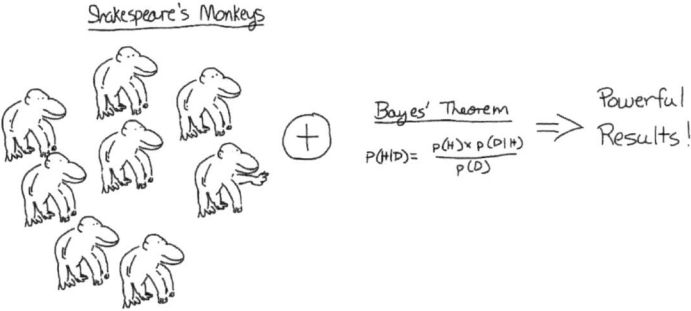

Now you have met the good Reverend Bayes and his incredibly useful formula. At this point you may be asking yourself, "How do you know what probabilities to put into the formula, and what happens if different people use different numbers for the same situation?" The first question can be answered by simply using your best guess, based on the information you know or have available. This may appear unsatisfactory, but using guided subjective thinking based on knowledge you know or can find will often lead to similar conclusions, even if the numbers are not exact.

If this sounds familiar to you, it should. This is simply a reiteration of the entire theme of this book: informed subjective estimates coupled with simple structured reasoning can produce surprisingly accurate models of the world. However, let me return to the question posed above and illustrate my point that Bayes' theorem is robust to these issues with the following hypothetical case.

Assume that we are attempting to determine whether a hypothesis (a diagnosis, for example) is likely true based on our subjective background knowledge. However, we are restricted to only using the following three probabilities: 25% (unlikely), 50% (no idea), and 75% (likely). I did not include the values of 0% and 100%, because both would imply absolute certainty, which is almost impossible to achieve (with the exception of death and taxes, as Benjamin Franklin famously wrote). We then restrict ourselves to only two interpretations of Bayes' formula: results ≤ 50% are equivalent to Not True, and results > 50% are equivalent to True. Thus, we are restricting ourselves to very crude inputs and outputs.

Now let's redo our analysis of the hypothesis regarding the patient with LOC due to a possible seizure, using our expanded formula:

$$p(H|E) = \frac{p(H|B)p(E|H)}{p(H|B)p(E|H) + (1 - p(H|B))p(E|\sim H)}$$

Previously I assumed that my prior was about 40%. Let's make things less certain and assume I am completely uncertain about the prior and thus use the 50% choice. My subjective probability of seeing a patient experiencing a seizure presenting with LOC was 90%. The largest probability choice I have is 75%. Finally, my earlier subjective probability for seeing a patient presenting with LOC who did *not* experience a seizure was 45%. Here my closest allowed probability choice is 50%. The results then become:

$$p(H|E) = \frac{0.50 \times 0.75}{(0.50 \times 0.75) + (0.50 \times 0.50)} = 0.60$$

Our results are not that far from the original values, even though we used cruder inputs to the equation. Our interpretation is also similar: the posterior probability is consistent with the hypothesis being true. The point of this example is to demonstrate that the Bayesian approach is flexible to inexact numbers, in that it will tend toward similar results even with variable inputs.

Now let's answer the second question on the previous page: What would happen if a number of different individuals looked at the same situation and attempted to assess the probability that a hypothesis is true, using Bayes' method with different inputs? Remember our problem with Shakespeare's monkeys? The famous "Infinite Monkey Theorem" suggests that an infinite number of monkeys banging on typewriters would eventually recreate the works of Shakespeare. We can use this concept and suppose that we have an unlimited supply of "virtual" monkeys that grab inputs from a defined range and plug them into Bayes' theorem. We then aggregate the results and see what we get. This type of exercise is known as a "Monte Carlo" simulation, based on the original method developed by Stanislaw Ulam during World War II.

Here we will change our original problem and randomly draw probability inputs from ranges. Let's define these as shown in the table below, where we now allow the original probability inputs to vary by ± 25% from the original, with the values clipped at 0% and 100%. Thus, we will use the following input ranges:

Variables	Min	Original	Max
$p(h|b) =$	15%	40%	65%
$p(e|h \wedge b) =$	65%	90%	100%
$p(e|{\sim}h \wedge b) =$	20%	45%	70%

I will assume that the inputs range between the values shown above, but I have no idea if these follow a specific distribution (beta, etc.). Therefore, I will allow my virtual monkeys to draw their inputs uniformly from the ranges shown above. For this simulation I ran 250 separate random draws and aggregated the results as shown below. They may surprise you—the final statistics are very close to those obtained in our original exercise:

POSTERIOR PROBABILITY

Mean	Median	5th% percentile	95% percentile
55.2%	54.2%	33.0%	75.8%

The simulation results show a mean value of 55%, instead of the 57% that my original estimates yielded. A key difference is that although this final result tends toward the earlier values, this method also indicates that the results cluster between 33% and 76%, depending on the actual inputs chosen. The following figure illustrates this graphically with the characteristic unimodal shape of the distribution being a result of "the law of large numbers."

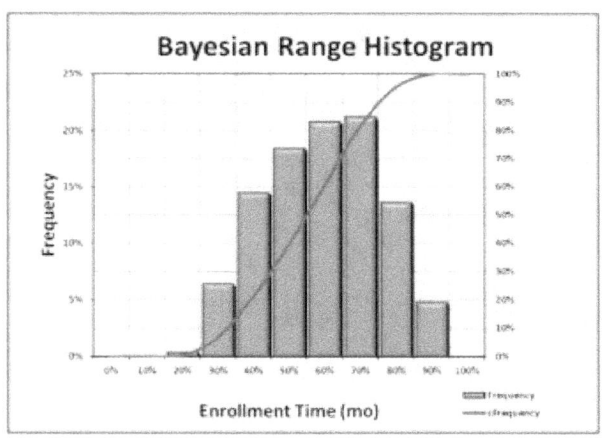

Figure 10. Histogram of Bayesian estimates

The cumulative distribution curve indicates that more than 50% of our "virtual monkeys" obtained results that suggest the hypothesis is true (the posterior probability is ≥ 50%). Thus, even using a range of uncertain inputs, the aggregate results are similar to our original analysis.

Hopefully I've shown you that the Bayesian approach is robust to varying uncertain inputs. Does this mean that Bayes is not subject to the "garbage-in-garbage-out" problem? No, the utility of Bayes' method still requires that you are using reasonably informed inputs—even if they are uncertain. If we redid the virtual-monkey simulation and now had all ranges vary from 0% to 100% uniformly, what would be the result? After 250 simulations, the median posterior probability result would be 50.7%, which is close enough to 50% to show that the results would be uncertain, and the 90th percentile of posterior probabilities would now range from about 8% to 94%. These numbers tell us that using completely uninformed "guessing" across the entire range of allowable probability values ends in an equally noisy range of results with the median and mean of approximately 50% (i.e., complete uncertainty).

The takeaway from this exercise is that Bayes' method is useful even in the face of uncertain inputs. However, the technique still relies upon the user to perform a reasonable level of due diligence in order to provide plausible inputs based on some understanding of the problem. In the case of healthcare problems, Bayes can be helpful if practitioners use their own baseline knowledge or aggregate input probabilities from subject-matter experts (physicians, nurses, etc.), depending on the problem.

CHAPTER 42

IS THIS DIAGNOSIS TRUE WITH NEW DATA?

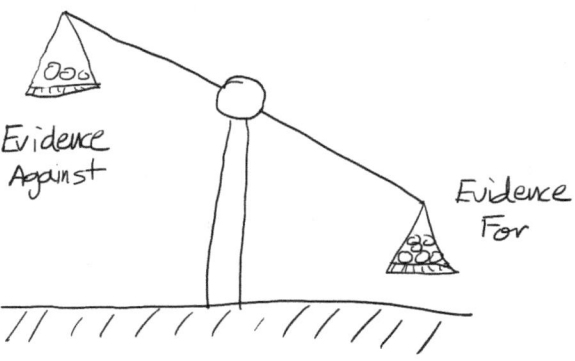

In our previous chapter, we used Bayes' theorem to evaluate the subjective probability that a hypothesis or conjecture is true based on background knowledge and one additional fact. What if we have several data items or facts to consider? It turns out that Bayes can be used multiple times to "update" our subjective probability based on new information, data, or facts. We will do that in this chapter by continuing our evaluation of the same patient with a presumed seizure. Here we use the same formula in an iterative fashion, using the previous posterior probability as our new prior, and turn the Bayesian crank to generate a new posterior probability based on the new information.

Note that as we move into multi-step Bayesian calculations, we are departing from the simpler, hand-calculator Fermi-type approach used in the majority of this book. Working through multi-step Bayes with a calculator and scrap paper begins to get very challenging, so I will resort to Excel to perform the calculations and put them into a table. Once again, the formula version we will use is:

$$p(H|E) = \frac{p(H|B)p(E|H)}{p(H|B)p(E|H) + (1 - p(H|B))p(E|\sim H)}$$

In terms of our clinical vignette, let's assume that based on the patient's history of a loss of consciousness (+LOC) we now think that the seizure hypothesis is more likely than not (57%, based on our calculations in the previous chapter). Now we begin our physical exam, and you notice that the patient has lacerations on his tongue or face and some bruising above the left orbital region. These physical signs can be associated with seizures and falls. We will call this a "constellation of signs" (+LAC) for tongue and/or face lacerations.

In general, observing these signs in patients experiencing a generalized tonic-clonic seizure (GTC) would be expected; we will therefore assume that our p(E|H) is on the order of 70%. In contrast, observing these signs in patients when they have not experienced a seizure would seem unlikely, and we will assume that our p(E|~H) is therefore somewhere around 15%. Note that p(E|~H) is not simply 100% minus p(E|H)! Our prior probability is simply the previous posterior probability of 57%. Thus, we can update our assessment as:

$$p(H|E) = \frac{0.57 \times 0.7}{(0.57 \times 0.7) + (0.43 \times 0.15)} = 0.86$$

At this stage, our probability that this individual patient suffered a seizure is 86%, based on our reasoning. Updating our assessment based on additional information can become tedious, so I will use a table to summarize our results so far:

MULTI-STEP BAYESIAN ESTIMATES FOR SINGLE HYPOTHESIS: PATIENT EXPERIENCED SEIZURE

Step	Data (evidence)	p(H|B)	p(E|H)	p(E|~H)	Num	Denom	p(H|E)
1	+LOC	0.40	0.90	0.45	0.36	0.63	0.57
2	+ LAC	0.57	0.70	0.15	0.40	0.46	0.86

You can see this new information has increased our probability from 57% to 86%. What if we add more information? Let's assume the patient also brought the results of two tests that looked for causes of an underlying seizure: an electroencephalogram (EEG) and a brain magnetic resonance image (MRI). Let's assume the EEG was unremarkable and did not reveal any obvious abnormalities in brain-wave rhythm consistent with a seizure disorder ("sharp" waves or notable asymmetries in brain activity). We call this a "negative" EEG.

In this case, we would expect that a patient who has suffered (or has a propensity for) seizures could have a single normal EEG approximately 60% of the time. The probability of a single normal EEG in a patient without seizures is even higher: approximately 80% of the time. (Note: The more EEG studies you obtain of an individual with an underlying seizure disorder, the higher the probability of obtaining at least one abnormal EEG.)

Rather than working this out manually, I've added the information into my Excel sheet to perform the Bayesian calculations for me. This new information actually reduces the probability of the patient having a seizure from 86% to 82%.

MULTI-STEP BAYESIAN ESTIMATES FOR SINGLE HYPOTHESIS: PATIENT EXPERIENCED SEIZURE

Step	Data (evidence)	p(H|B)	p(E|H)	p(E|~H)	Num	Denom	p(H|E)
1	+LOC	0.40	0.90	0.45	0.36	0.63	0.57
2	+Tongue laceration	0.57	0.70	0.15	0.40	0.46	0.86
3	-EEG test	0.86	0.60	0.80	0.52	0.63	0.82

Finally, we look at the patient's brain MRI to see if it adds any further information. To make this interesting, I assume that the MRI is abnormal and shows left temporal sclerosis (scarring), which is a relatively common abnormality seen with seizure disorders. Knowing this abnormality, I estimate that the probability of observing this in a patient with a propensity for seizures is approximately 85%. However, not all patients with this abnormality experience seizures; the probability of this abnormality occurring in patients without seizures is about 40%. Putting this information into my Bayesian calculator, the probability that the patient experienced a seizure increases from 82% to 91%.

MULTI-STEP BAYESIAN ESTIMATES FOR SINGLE HYPOTHESIS: PATIENT EXPERIENCED SEIZURE

Step	Data (evidence)	p(H|B)	p(E|H)	p(E|~H)	Num	Denom	p(H|E)
1	+LOC	0.40	0.90	0.45	0.36	0.63	0.57
2	+Tongue lac	0.57	0.70	0.15	0.40	0.46	0.86
3	-EEG test	0.86	0.60	0.80	0.52	0.63	0.82
4	Abnormal MRI	0.82	0.85	0.40	0.70	0.77	0.91

In summary, with this clinical vignette we have updated our initial probability for our seizure hypothesis from 57% to 91%, based on several successive pieces of information. This illustrates the utility of Bayes' theorem for modeling clinical decision-making.

CHAPTER 43

WHICH DIAGNOSIS IS MOST LIKELY?

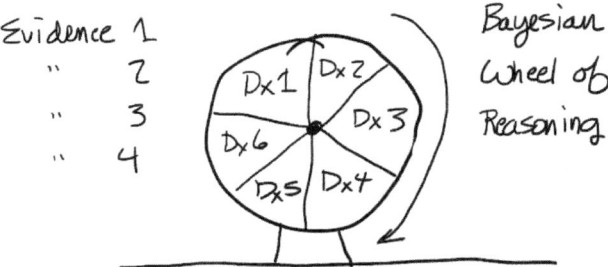

Until now, we have used Bayes' theorem with data to determine our subjective probability or belief that a single hypothesis or conjecture is true. As physicians, and in life, we often have more than a single possibility to consider. My final use of Bayes in this book will be to work through a hypothetical example where a patient presents with a constellation of data and we must determine which of several hypotheses are most consistent with the facts. While this may appear to be a new application, it is simply extending our previous work to more than a single possible hypothesis.

As with our previous multi-step Bayesian calculations, we have departed from the hand calculator Fermi-type approach, so again I will resort to Excel to perform the calculations and put these into a table to illustrate the numbers. This time, I will rewrite our Bayesian formula version differently to aid the calculations:

$$p(H|E) = \frac{p(H|B)p(E|H)}{p(H|B)p(E|H) + (1 - p(H|B))p(E|{\sim}H)}$$

$$= \frac{p(H|B)p(E|H)}{p^*} = \frac{prior \times likelihood}{normalizing\ constant}$$

What gives? Here I have condensed the denominator into a single factor that is actually disregarded in many Bayes calculations, or used as a value to make our calculations reasonable. This term is known as "the marginal likelihood" (also called a "normalizing constant") and does not have any effect on which hypothesis will yield the largest posterior (probability), so we will not consider it in detail. However, it will be important to calculate it in order to keep our individual posterior probabilities well behaved and interpretable.

When evaluating multiple hypotheses at once, we need to have an exhaustive list, or at least one that includes the majority of possibilities. Otherwise, we can find that our analysis is biased simply because we limited our hypotheses to only a few.

With this in mind, let me set up our clinical scenario. Imagine you are called into an urgent care setting to evaluate a young woman with a headache (HA). You consider four possible causes, two of which are potentially malignant. To illustrate the utility of the Bayesian method, I will make my priors all naïve (also referred to as "an uninformed prior"), with the sum equaling one (i.e., each potential diagnosis has the same prior probability). In real life, I know that a migraine would be the most likely cause in this population, and I would adjust the priors accordingly. However, let's make our initial estimate 25% for each prior. Next, I must estimate the likelihood that I would see the evidence (first piece of data is that the HA is unilateral) if the hypothesis is true.

The following table uses the approach described by aerospace engineer Scott Hartshorn.[48] It illustrates the naïve priors for our initial probabilities and my assessments of the likelihood of the HA being unilateral if each of the hypotheses were true. Note that I guessed at these numbers based on my background as a neurologist; one could also estimate them from epidemiologic data. Also, I have judged 75% probability for observing a unilateral HA with migraines, a 20% chance with Tension HA, and 10% each for the malignant conditions of subarachnoid hemorrhage (SAH) and meningitis (MGT). In contrast to our previous problem, I obtain the marginal likelihood by summing the p* values (not to be confused with the term *p-value* from statistical hypothesis testing) across the hypotheses and then dividing each by the total. This is called the "normalizing constant" and is the denominator for Bayes' equation. Using this gives me my posterior probability:

MULTI-HYPOTHESES FOR HEADACHE:
UNILATERAL HEADACHE

Hypothesis	p(H|B)	p(E|H)	p*	p(H|E)
Migraine headache (MHA)	0.25	0.75	0.19	0.65
Tension headache (THA)	0.25	0.20	0.05	0.17
Subarachnoid hemorrhage (SAH)	0.25	0.10	0.03	0.09
Meningitis (MGT)	0.25	0.10	0.03	0.09
Sum	1.00	1.15	0.29	1.00

The above table shows that given the first set of data, our most likely hypothesis is a migraine HA (MHA). Next, we update our estimates with additional observations provided by the patient's history, exam, and testing. Let's assume we observe on examination that our young patient has nuchal rigidity (stiff neck). In my experience as a neurologist, and that of most physicians, this is known to be one of the signs of meningeal infections and is not consistent with migraine or tension

---

[48] Scott Hartshorn, *Bayes' Theorem Examples: A Visual Guide for Beginners*, 2016, Kindle edition.

HA syndromes. In this set, I estimate that the likelihood of observing a patient with nuchal rigidity and having an MHA, tension HA, subarachnoid hemorrhage, and meningitis as 15%, 15%, 35%, and 75% respectively. Finally, we update our Bayesian estimates using the posterior results as our new priors and multiplying these with our likelihoods of observing stiff neck with the various hypotheses:

MULTI-HYPOTHESES BAYESIAN ESTIMATES FOR HEADACHE: STIFF NECK

Hypothesis	p(H|B)	p(E|H)	p*	p(H|E)
Migraine headache (MHA)	0.65	0.15	0.10	0.45
Tension headache (THA)	0.17	0.15	0.03	0.12
Subarachnoid hemorrhage (SAH)	0.09	0.35	0.03	0.14
Meningitis (MGT)	0.09	0.75	0.07	0.30
Sum	1.00	1.40	0.22	1.00

Here I made the likelihood much higher for observing nuchal rigidity with meningitis. Consequently, the resulting posteriors decrease the probability of migraine and increase the probability of meningitis. As we discussed in Chapter 41 on Bayes and Monte Carlo simulation, as long as the likelihood percentages (or proportions, as I write them in the table) are roughly similar to mine, the overall results will tend to be similar as well. You might note that the individual likelihood values p(E|H) do not sum to 1.0. This is perfectly OK. The individual likelihoods are independent estimates without regard to the other hypotheses and are not probability distributions in themselves, and as a result will not sum to 1 or 100%. In contrast, both the priors and the posteriors must sum to 1.0.

Currently our grid of hypotheses suggests that migraine is still most likely but decreasing in probability, while meningitis has greatly increased from less than 10% to 30%. Next, we will assume that the patient is also observed to have an above-normal body temperature (fever) associated with the unilateral headache and stiff neck. Again, we must estimate our likelihoods for the different hypotheses and then turn the Bayesian crank to generate new posterior probabilities for our potential diagnoses.

From my personal experience in practice, as well as the medical literature, fever is not a typical finding with migraines or tension-type headaches. It is also not common with SAH, although it is potentially more frequent than with the primary headache syndromes. However, like stiff neck, fever would not be unusual at all with meningitis. For this reason, I estimate the subjective likelihood for my different diagnostic hypotheses as 10%, 10%, 20%, and 80%, respectively. As discussed previously, if you come up with different specific values, as long as these are reasonably close to my own (within ±20 or 25%), you will come out with a hypothesis ranking that is roughly similar to the results in the table below.

MULTI-HYPOTHESES BAYESIAN ESTIMATES FOR HEADACHE: FEVER

Hypothesis	p(H|B)	p(E|H)	p*	p(H|E)
Migraine headache (MHA)	0.45	0.10	0.04	0.14
Tension headache (THA)	0.12	0.10	0.01	0.04
Subarachnoid hemorrhage (SAH)	0.14	0.20	0.03	0.09
Meningitis (MGT)	0.30	0.80	0.24	0.74
Sum	1.00	1.20	0.32	1.00

As you can see from our calculation table, meningitis is now our most likely diagnosis, followed by migraine, SAH, and tension HA. In order to proceed with a diagnosis in a case like this, a doctor must look at brain imaging to rule out hemorrhage (consistent with SAH) or mass before proceeding with further invasive evaluation, such as a lumbar puncture to obtain cerebral spinal fluid. Let's assume that a head CT has been performed that appears normal. A normal head CT would typically be expected with migraine, tension HA, and meningitis, but not with SAH. Therefore, I estimate the likelihoods as 80%, 90%, 15%, and 80%, respectively. When we perform our calculations, our posteriors still show meningitis as the most likely diagnosis, with migraine a faraway second.

Multi-Hypotheses Bayesian Estimates for Headache: Normal Head CT Scan

Hypothesis	p(H|B)	p(E|H)	p*	p(H|E)
Migraine headache (MHA)	0.14	0.80	0.11	0.15
Tension headache (THA)	0.04	0.90	0.03	0.04
Subarachnoid hemorrhage (SAH)	0.09	0.15	0.01	0.02
Meningitis (MGT)	0.74	0.80	0.59	0.79
Sum	1.00	2.65	0.75	1.00

As a final step, I will assume that with the normal head CT, a lumbar puncture (LP) is then performed. Note that in actuality, the presence of headache, stiff neck, and fever are the typical triad of meningitis, and in many cases, presumptive treatment would have already been initiated. Let's assume that the results of our LP show increased pressure and cloudy fluid. I would then estimate the likelihood probabilities as 10%, 10%, 20%, and 80%, respectively. This gives us our final posterior values of:

Multi-Hypothesis Bayesian Estimates for Headache: LP with Increased Pressure

Hypothesis	p(H|B)	p(E|H)	p*	p(H|E)
Migraine headache (MHA)	0.15	0.10	0.01	0.02
Tension headache (THA)	0.04	0.10	0.00	0.01
Subarachnoid hemorrhage (SAH)	0.02	0.20	0.00	0.01
Meningitis (MGT)	0.79	0.80	0.63	0.97
Sum	1.00	1.20	0.65	1.00

In this clinical scenario, we have begun with naïve priors for our potential diagnoses and ended up with a most likely diagnosis of meningitis based on a series of tests. This is illustrated in the graph below. (After putting the numbers in such nice, clean tables, I could not give you one of my hand-drawn graphs of the results!)

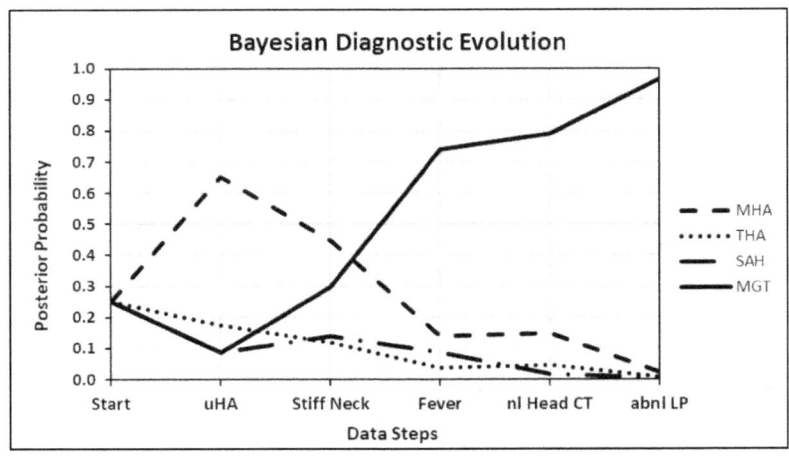

Figure 11. Bayesian posterior graph

In real life, no one would go through such a cumbersome exercise when attempting to diagnose a patient in an emergency. However, the Bayesian approach does a nice job of explicitly illustrating an analog of the diagnostic process we employ as healthcare providers. Physicians, nurse practitioners, and physician assistants learn to do this organically through experience, without reference to tables or explicit algorithms. However, with the rise of machine learning, vast data, and inexpensive processing, there's hope that smart-decision support systems can be developed to augment the work of our clinical diagnosticians. You can be sure that Bayesian methods will be used as part of this effort!

CHAPTER 44

OH ERROR, WHY ART THOU? THE RELIABILITY OF PROCESSES

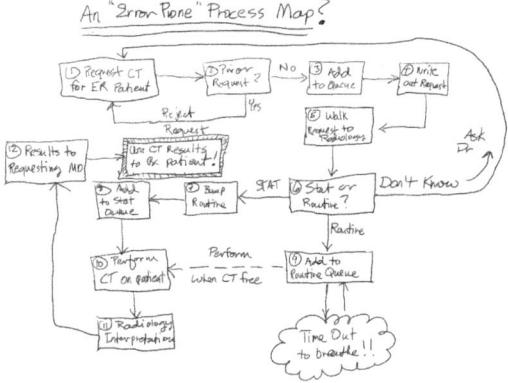

Medication errors, missing or wrong product parts, and late deliveries unfortunately do occur in our modern world. Within healthcare, adverse drug events and medication errors are a significant problem, costing the health system approximately $3.5 billion in 2016.[49] While the root cause of these types of events can be varied (medication order not received, wrong dosing frequency, etc.), I want to explore how the process flow itself can lead to unanticipated problems, even when you have individuals trying to do the right thing.

I will use hospital medication errors to illustrate the mathematics of

---

[49] Jay Arthur, *Lean Six Sigma for Hospitals,* 2nd ed. (New York City: McGraw-Hill Education, 2016).

the reliability of a process in general. This same approach can be used to evaluate failure rates for integrated circuits, misses for a pharmacovigilance system, or defect rates in a manufacturing line.[50] Unfortunately, I am not aware of any published examples in which this analysis has been performed in a healthcare situation.[51]

Let's begin by examining a typical manual activity flow for a doctor who wants to prescribe a particular anti-epileptic drug for a patient. This is summarized in the figure on the next page, where the doctor first determines the medication, dose, and frequency to administer the medication. The doctor then enters the prescription order by writing it or using a computerized order-entry system. In some situations, these orders are still manually transcribed from one system to another and the medication is requested. As a final step, the nurse receives the medication and administers it to the patient. Note that there can be variations in these steps, and there is currently much more use of computerized entry, bar codes, and so on that can substantially improve the cumulative error rate these manual processes can produce. However, here we will assume that these processes are all manual.

Before getting to the mathematics here, it is important to note that "error" can mean different things: sometimes a mistake has clinical consequences (wrong drug, wrong dose, wrong route), and sometimes the mistake has no observable clinical effect (delay, order not received, sent to wrong floor, near-miss with wrong medication). These latter nonclinical consequential errors still cost the system in lost time, delayed treatment, and staff rework, as well as the opportunity cost of being unable to treat other patients while mitigating errors. For our purposes, I will use the generic term "error" to refer to all these examples.

---

[50] D. A. Marx and A. D. Slonim, "Assessing Patient Safety Risk Before the Injury Occurs: An Introduction to Sociotechnical Probabilistic Risk Modelling in Health Care," abstract, *Quality & Safety in Health Care* 12, Supp. 2 (December 2003): ii33–ii38, doi:10.1136/qhc.12.suppl_2.ii33.

[51] John Wreathall and C. Nemeth, "Assessing Risk: The Role of Probabilistic Risk Assessment (PRA) in Patient Safety Improvement," abstract, *Quality and Safety in Health Care* 13, no. 3 (2004): 206–212, doi:10.1136/qhc.13.3.206.

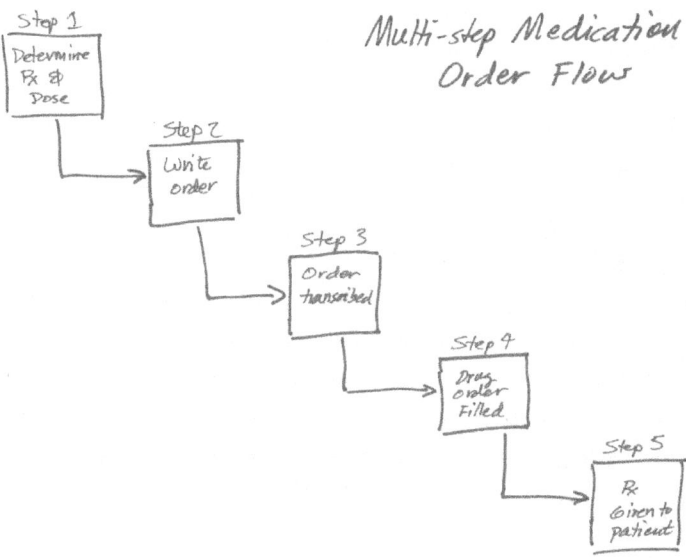

Figure 12. Hypothetical medication order flow

Assume that for each of the steps in the figure, there is an associated reliability (or "non-error") rate. For example, let's assume that our doctor is able to determine the appropriate medication and dose with 99% reliability (a 1% error rate). The doctor performs each of the subsequent steps with similar 99% reliability. If we treat step reliability as a probability of no error, we can then determine the aggregate error for the entire process. If we further assume that each step is independent of the other (determining dose doesn't affect transcription, for example), we can estimate the overall reliability as follows from the rules of probability:

$$p(A \cap B \cap C) = p(A) \times p(B) \times p(C)$$

This is an imperfect assumption, because if the doctor chooses the wrong drug, the error could have a meaningful clinical consequence, even if all the subsequent steps are performed perfectly. However, our figure is a reasonably conservative assumption that will illustrate the concept. For our particular case, we can rewrite the aggregate process reliability (probability of no error) as:

$$P_r = p(A) \times p(B) \times p(C) \ldots = \prod_{i=1}^{n} P_i = 0.99^5 = 0.95 \text{ or } 95\%$$

Thus, our process has 95% reliability, even though each of the components operates at 99% reliability. It's important to note that the aggregate reliability will always be less than the best component reliability. In our case, this becomes approximately 50 errors per 1,000 orders. Is 99% reliability realistic for humans? Based on the medical literature and studies on accuracy by clinical psychologist Paul E. Meehl and others, human performance on diagnostics range between 60–90% accuracy.[52] Assuming this is representative of average human performance with subject-matter expertise and training, linear processes like this would actually perform much worse. For example, if we assume that each component is performed at 95% reliability, our results become:

$$P_r = \prod_{i=1}^{n} P_i = 0.95^5 = 0.77 \text{ or } 77\%$$

In an academic setting, achieving 95% accuracy on a test would be considered excellent work. However, when this performance is linked together, we have only a 77% reliability and a 23% error rate! These results may be worrisome, but again we should remember that many of these errors might have no clinical consequences and could be truly minor (e.g., a one-letter spelling error on a transcription that doesn't affect the order). Is there anything that can be done to improve the accuracy? Yes, we could build in some parallel processes and checks to improve things. I've illustrated one hypothetical improvement below in a wire diagram:

---

[52] Shim and Siegel, *Operations Management*. (See full cite in footnote 26.)

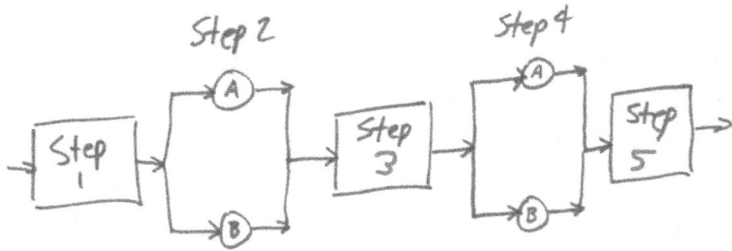

Figure 13. Hypothetical parallel process flow

Here I have introduced parallel processes for Steps 2 and 4, so that if an error occurs in one arm of the parallel process, it can be captured or corrected via the other arm. One example from data management is the use of double-data entry, where two independent writers enter information into a database and the results are compared to detect errors and correct them before finalizing. Another word to describe this is "redundancy." In critical systems, backups and parallel features ensure quality and reliability.

Note that introducing parallel steps changes both the calculations and results, because for the parallel processes we calculate the reliability as:

$$P_r = 1 - \left[\prod_{i=1}^{n} (1 - P_i) \right]$$

Now let's redo the order entry calculations, with each step at 95% reliability. Here we can see that we achieve an overall result of:

$$P_r = P_1 \times \left(1 - (1 - P_{2a}) \times (1 - P_{2b})\right) \times P_3 \times \left(1 - (1 - P_{4a}) \times (1 - P_{4b})\right) \times P_5 = 0.854 \approx 85\%$$

Adding in two parallel processes increases the reliability to 85% and reduces the error rate from 23% to approximately 15% (a 35% decrease). If we go back to our 99% component-step reliability, the results are even better.

We can draw some obvious conclusions from this simple example: long serial processes can introduce large cumulative-error rates, even when individual steps operate at the upper limit of human performance.

Here we are talking about errors that can accumulate through the process. In some cases, these errors can stop a process (an electronic circuit, machine failure on a manufacturing line, etc.), and it is important to think through the weaknesses of processes to improve the reliability for critical activities. Computerized order entry for medication and other tools such as checklists have made some improvements in healthcare.

So how can we achieve better reliability? We have three primary choices:

1. Demand near-perfect reliability for each serial step (typical proposal but very hard to do, especially with manual activities)

2. Minimize process steps (keep things simple)

3. Incorporate parallel steps within the most critical serial parts of the process (can be expensive but makes a significant difference)

In summary, I hope you can now see how mathematics shows that many of our very complicated steps in medicine and healthcare can lead to significant issues. It seems that the more steps and routes we have in our processes, there are more ways to make mistakes. (Although this truism is intuitively known, it's generally unrecognized in the healthcare industry.)

The corollary in physics and engineering is the concept of entropy: complex systems tend to become more disordered. We acknowledge this reality in our everyday lives with sayings like: "Anything that can go wrong, will go wrong," "Situation normal: all fouled up" (SNAFU), and "Keep it simple, stupid!" Entropy is one of the most insightful and powerful concepts in physics and is fundamental to how our industrial technology works, how bodies and machines age and fail, why ice cubes melt, and why errors propagate.

CHAPTER 45

WHY PROJECTS (AND ALMOST EVERYTHING ELSE) RUN LATE

I magine that you are the team leader for developing a study protocol (or piece of software, medical device prototype, hospital budget plan, etc.) and are working with five team members. To move forward with the next phase, you need all members to review and provide edits to your draft protocol, with each reviewer taking between five and nine days to perform this. In discussion with your team, all agree that an average of seven days is required for each to complete their work and return for the project manager to compile. What is the probability that this will be done by ≤7 days?

Figure 14. Gambling project manager

Here we are given the average of seven with a maximum range of five to nine days. Given the symmetrical nature of the numbers, I will assume we are dealing with a normal Gaussian (aka "normal") curve instead of a log-normal curve, and estimate the standard deviation using the PERT formula with the range of completion days as:

$$s_d = \frac{(9-5)}{6} \approx 0.7 \, days$$

By assuming that a normal curve is a reasonable first approximation, this has the advantage of making the math a little simpler for illustration purposes. For a Gaussian distribution, the mean and the median (50th percentile) are equivalent, with one half of the distribution below these values and one half above.

Let's assume we are only interested in whether the team reviews occur earlier or later than the seven-day median. This allows us to treat the situation as discrete binomial draws or flips of fair coins where heads (H) corresponds to an individual reviewer completing their task at ≤ 7 days (the 50th percentile) and tails (T) corresponds to completion at > 7 days.

How Much Is that Cure in the Window?   171

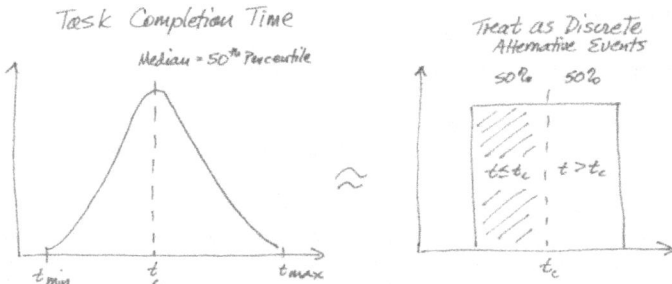

Figure 15. Treating completion time as binary event

The desired outcome is for all five reviewers to complete their tasks within the seven-day median. What is the probability that this will occur? Given that each review is assumed to be independent, we can use the laws of probability to calculate the result as the multiplication of each probability:

$$P = p(r_1 \leq 7) \times p(r_2 \leq 7) \times \ldots \times p(r_5 \leq 7) = p(H)^5 = 0.5^5 \approx 3\%$$

Wow, this is not an encouraging outcome! Since the project cannot move forward until all the reviews are completed, we can see that the probability of uncompleted reviews by day seven is around 97%. Why is this the case? Here is where the "coins" may help understand what is happening. In order to meet our goal, all five independent reviewers must complete their reviews at their median time. Basically, we are flipping five independent fair coins and expecting them all to come up heads! It is much more likely that with these five coins you will get a combination of heads and tails.

Using Hartshorn's formula for combinatorics, we can determine that there is only one combination of ways where these five coins all come up as heads:

$$nC(5,5) = \frac{N!}{n!\,(N-n)!} = \frac{5 \times 4 \times 3 \times 2 \times 1}{5 \times 4 \times 3 \times 2 \times 1} = \frac{5}{5} = 1$$

In contrast, there are 32 total possible combinations (2^5), only one of which (five heads) gives us the desired outcome we want. This illustrates an important point about why so many projects are late: because there are so many more ways to be late than to be on time (or early).

Let's return from the world of coins and ask ourselves what duration of time would be required to be 90% confident all the reviews will be completed. Here we first note that the distributions are the same for each reviewer, and determining the 90th percentile for any reviewer will be the same as the 90th percentile for the group. We can use our assumption of distribution normality to our favor by drawing on the formula for a standard normal curve known as z. This statistic is defined as follows:

$$z = \frac{x - \mu}{\sigma}$$

Since z values are calculated for typical values such as 90%, we can solve for the time corresponding to the 90% value of z, which is 1.65. This gives us our estimate for the 90% confidence limit (or percentile) for completion time:

$$= \mu + z\sigma \equiv 7 \; days + 1.65 \times 0.7 \; days \approx 8 \; days$$

We could also calculate an estimate for the 99th percentile for completion time. However, I think we have made the point that it is highly unlikely that all independent tasks will be completed by their individual average times.

What happens with sequential tasks? What if our scenario consisted of five independent tasks that must each be completed before the next begins, instead of all finishing together? Let's work this out and see. To make it more realistic, let's make the units in weeks instead of days, but keep the other numbers the same.

First, we can again use the fair-coin scenario, and whether we flip five independent coins together or sequentially does not alter the results. Thus, we can immediately say that the probability of having all tasks completed at their median times will again be about 3%. Note that here we add the medians to get our overall expected time of completion of 35 weeks (7 × 5). However, the overall distribution for completion times will be larger than for the identical subtasks, unlike our earlier example. This comes from the discussion earlier in the book where uncertainty propagates when adding, subtracting, etc. As a reminder, the overall standard deviation is calculated as follows and can be simplified in our case because the individual standard deviations are equivalent:

$$\sigma \approx 2.2 \times 0.7 = 1.6$$

Similar to our previous example, we can calculate the completion time corresponding to our 90% and 99% upper confidence limits as:

$$= \mu + z\sigma = 35 \text{ weeks} + 1.65 \times 1.6 \text{ weeks} \approx 38 \text{ weeks}$$

$$= \mu + z\sigma = 35 \text{ weeks} + 2.6 \times 1.6 \text{ weeks} \approx 39 \text{ weeks}$$

Notice that the combination of independent identical stochastic events produces a widened aggregate distribution relative to the individual subtasks. The result is that our project's 90th-percentile completion time is over three weeks late, with a 10% chance that completion will be even later. Why does this happen?

One key reason is that we are using means (here equivalent to medians, in our case of the Gaussian distribution times) derived from stochastic processes (chance-based processes) as inputs for deterministic schedules and plans. Sam Savage[53] and Patrick Leach[54] treat the significant limitations of this common approach both humorously and mathematically, in more detail than I do here.

A primary problem with simply combining single-point estimates when they are really expectations from distributions is that we are in effect hiding the uncertainty that is inherent in the world. A project timeline obtained from five tasks performed in parallel, or series, is just one desired sample path out of many possible ones allowed by the laws of chance. Ignoring the distributional aspect of these estimates almost guarantees the common outcome of missing timelines and being late.

---

[53] Sam L. Savage, *The Flaw of Averages: Why We Underestimate Risk in the Face of Uncertainty* (Hoboken, NJ: Wiley-Blackwell, 2012).

[54] Patrick Leach, *Why Can't You Just Give Me the Number? An Executive's Guide to Probabilistic Thinking to Manage Risk and to Make Better Decisions*, 2nd ed. (Sugar Land, TX: Probabilistic Publishing, 2014).

What can be done to improve this? As Savage and Leach point out, creating stochastic timelines and using Monte Carlo simulations can go a long way to making project plans more realistic. However, this is almost universally not done in all the areas of healthcare I am familiar with.

In contrast, the usual approach I have witnessed in building budgets, research plans, and timing models is to "pad" the results with a fudge factor based on gut instinct, rules of thumb, etc. This common tactic at least indicates that someone recognizes the problem of underestimating time and budgets, but adding in silly factors based on instinct are not the answer and introduce their own problems with consistency.

Note that the examples shown here all assumed Gaussian-normal completion-time behavior. In fact, it is well known that task times typically are highly skewed and are better represented by beta or log-normal distributions.

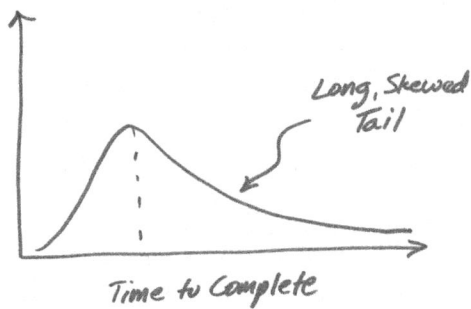

Figure 16. Skewed completion-time distribution

The result is that using normal distribution assumptions will further bias our time estimates toward overly optimistic results and increase the likelihood of time delays, because there are real physical limits on how quickly a task can performed (how fast a chart can be read, how rapidly a patient can be seen in clinic, etc.). In contrast, there are few limits on how long it can take.

In summary, I hope you can see that based on the math alone, almost all projects and activities of any meaningful complexity are likely to take longer than expected. Given a world where most of the population is innumerate and probability-blind, very few individuals are aware of these issues. The result is that most of our business and life venture plans are overly optimistic and flawed. A good rule of thumb is that when you view a project timeline, a budget estimate, or a cost model, it's likely that the project will end up late, the budget will be inadequate, and the costs will be more than anticipated. No wonder so many business ventures fail!

CHAPTER 46

TYING THINGS TOGETHER: FROM FERMI ESTIMATION TO ENTROPY

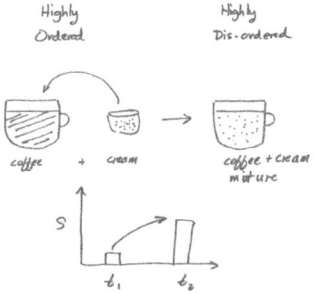

We began our journey by using "Fermi estimation" to approximate quantities of interest to an order-of-magnitude precision. We progressed from relatively simple problems to more complicated ones, like estimating health insurance and Alzheimer's care costs. We then moved into using Bayesian thinking to assess how likely a clinical diagnosis was, based upon various data. Finally, we explored some of the implications for how reliable processes can be and why our plans and projects are so often late. In this last chapter, I want to build upon my previous introduction to entropy, provide some additional background on its history and importance, and finally link this concept to the Fermi estimation method we have used throughout this book. Let's begin!

In the late nineteenth century, the Industrial Revolution was well underway, with steam-powered machines replacing biological muscle as the primary means of performing work. Our ability to create engines that do mechanical work is only possible due to the underlying laws of thermodynamics, which govern how energy is transformed from useful states that can drive crankshafts, beat hearts, or allow brains to compute, to unusable heat and disorder. The second law of thermodynamics states that for any closed system such as a cup of coffee with cream carefully poured into a discrete layer on top, the system will tend toward more disorder—in this case, the cream and coffee will become mixed. The result is that this disordered system will not spontaneously go back to its original highly ordered state without the input of energy to separate the individual components back to their original state.

During this time, the late 1800s, the brilliant theoretical physicist Ludwig Boltzmann derived the underlying equations that prove the "second law" (as it is affectionately known in physics and engineering), a statistical law where any system such as a mixture of gases, fluids, etc., will move to its most probable state.[55] The contributions of Boltzmann and his contemporaries (Gibbs, Maxwell, and others) to various branches of physics are legendary, and I encourage anyone interested to delve into the rich literature of their history. I am only giving a very superficial overview here.

To bring this back to our example and this chapter, the most probable visible "macro" state a system will be in (coffee and cream mixture, a system of flipped coins, a complicated machine, a healthcare system) will be determined by how many possible ways events can occur, unless additional work is performed to put it in an ordered state. Another way to say this is that everything in the universe naturally tends toward states of greater disorder, from stars and animals to liquids and our own bodies. This occurs because there are so many more ways for systems

---

[55] David Lindley, *Boltzmann's Atom: The Great Debate That Launched a Revolution in Physics* (New York: Free Press, 2016).

or plans to be disordered than ordered.

To make this more tangible, let's return to our five-coin project management example from Chapter 45, where there were 32 ways for the system to be late, but only one possible way it could occur on time. Therefore, the chance of the project achieving its time target was only 3%, but in contrast, had a 97% chance of being late. Professor Boltzmann developed an equation that concludes that entropy is proportional to the number of ways that a system can be in something called "microstates."[56] I've written the formal equation below, as well as a second formula for the number of possible heads/tails combinations in a five-coin system:

$$S = kLn\Omega$$

$$\text{where } \Omega = \frac{N!}{n!(N-n)!}$$

In our coin-flipping example, the constant "k" is set to one. As you will see in the table below, calculating out the values for our coin project shows that the highest entropy corresponds to the states with the greatest number of possible ways of occurring (highest probability), whereas the desired state (everyone finishing at their stated time, all coins coming up heads) has a very low entropy. A trivial way to state Boltzmann's law is by noting that a system will tend to be in its most probable state.

ENTROPY FOR A SYSTEM OF 5 FAIR COINS

Heads	Tails	Ω (# of combinations)	Prob	Entropy (S)
5	0	1	3%	0.00
4	1	5	16%	1.61
3	2	10	31%	2.30
2	3	10	31%	2.30

---

[56] Stephen J. Blundell and Katherine M. Blundell, *Concepts in Thermal Physics*, 2nd ed. (Oxford, England: Oxford University Press, 2009).

1	4	5	16%	1.61
0	5	1	3%	0.00
Totals		32	100%	

OK, you might say. So what? How is this related to missing timelines, the information in a book, doctors running late, health plans busting budgets, and clinical trials under-enrolling? All these different problems are related because they are all manifestations of the same underlying issue: stochastic unfolding of complicated, interacting processes that will always tend toward maximum-entropy states, because they are the most probable.

I think many colloquial sayings like "measure three times and cut once" have their origin in our recognition of how entropy naturally increases in the world around us. One of my favorite clinical operations directors once said, "Shaun, you need to have eyes in the back of your head all the time, or the program will go off the rails."

However, entropy is not always a negative. The universal tendency for disorder or entropy to spontaneously increase is what allows heat to only flow from warm bodies to cold ones, and gives rise to the irreversible thermodynamic arrow of time we experience every day.[57] Entropy and thermodynamics are also closely related to information entropy and heat production as a byproduct of computation and information storage.

So how does this relate to Fermi estimation? If you remember from Chapter 5, I said that one could view Fermi ranges as subjective "confidence intervals" that could be designed to be 90% likely to contain the desired quantity. Using this interpretation and combining that with the insights from Chapter 6 on Uncertainty Propagation, we are in the position to make a plausible argument for why Fermi estimates work so well, often achieving much better results than the advertised "order of magnitude" precision.

---

[57] See, e.g., Andrew Thomas, *Hidden in Plain Sight 3: The Secret of Time* (self-published, 2014) and Sean Carroll, *The Big Picture: On the Origins of Life, Meaning, and the Universe Itself* (Boston: Dutton, 2017).

So why is that? This technique likely works for several reasons. One is that breaking problems into smaller sub-problems allows us to make more precise estimates than simply generating one grand guess. Second, we can usually assume that each estimate's error is independent of the others in a problem. Therefore, when we combine the point estimates together in a calculation, the over-estimates and under-estimates will tend to cancel each other out.

I speculate that this statistical tendency to get closer than expected to the correct answer for linear Fermi problems is related to something known as "the maximum entropy principle," where the probability distribution best representing the current state of one's knowledge has the largest entropy. However, I cannot prove that, and my conjecture could turn out to be incorrect.

The sum of the general principles I've outlined is this: We can and should always do our best to forecast, make thoughtful decisions, estimate quantities, troubleshoot, and manage our plans and hopes, but we only have so much control and ability to influence life. The practice of medicine illustrates this point well, as unpleasant surprises and the unexpected can occur in spite of the best intentions, treatment plans, and technology. Anyone who receives an unexpected diagnosis of cancer at a routine doctor visit, such as I did, will appreciate this point clearly.

Finally, below are four points I hope you take away from this book:

1. Naïve plans that ignore uncertainty and stochasticity are the norm in medicine, healthcare delivery, and much of life. The fact that the majority of plans built on these techniques fail should be expected.

2. Making simple, clearly defined models and techniques that incorporate uncertainty in the inputs and are clearly communicated to the user can go a long way to supporting better decision-making and outcomes. I suggest pursuing Fermi models first to evaluate overall feasibility before launching into major

programs and decisions.

3. Maintaining a humble attitude while developing your skills as a good approximator and modeler is a good way to minimize fooling yourself.

4. The problems we deal with in the healthcare field are deeply embedded in the way the universe works, just like in every other field. Entropy and its trend towards increasing disorder are fundamental. Recognizing this fact from physics and incorporating this insight and techniques from other fields into our toolkits can help us focus more on realistic, achievable goals in medicine.

In conclusion, I want to thank you for your time and interest in this book. It is the culmination of 30 years of notes and ideas scribbled on pages and journals. I hope you enjoyed the journey!

ADDITIONAL REFERENCES

Alemi, Farrokh, and David H. Gustafson. *Decision Analysis for Healthcare Managers.* Chicago: Health Administration Press, 2006.

Edwards, D., and M. Hamson. *Guide to Mathematical Modeling.* 2nd ed. London: Palgrave Macmillan, 2007.

Francis, Paul. "Back-of-the-Envelope Calculations: Or: The Seven Habits of Highly Effective Astronomers." Australian National University Research School of Astronomy and Astrophysics website, February 25, 1999. http://www.mso.anu.edu.au/pfrancis/Approximations.pdf.

Gorini, Catherine. *Master Math Probability: Master Everything from Sample Spaces and Counting Techniques to Expected Value and Laws of Large Numbers.* Boston: Course Technology PTR, 2011.

Grove, W. M., D. H. Zaid, B. S. Lebow, B. E. Snitz, and C. Nelson. Abstract. "Clinical Versus Mechanical Prediction: A Meta-Analysis." *Psychological Assessment* 12(1) (March 2000): 19–30.

Harte, John. *Consider a Spherical Cow: A Course in Environmental Problem Solving.* Mill Valley, CA: University Science Books, 1988.

———. *Consider a Cylindrical Cow: More Adventures in Environmental Problem Solving.* Mill Valley, CA: University Science Books, 2001.

Lyons, Melinda, Sally Adams, Maria Woloshynowych, and Charles Vincent. "Human Reliability Analysis in Healthcare: A Review of Techniques." *International Journal of Risk & Safety in Medicine* 16 (2004): 223–237.

Swartz, Clifford. *Back-of-the-Envelope Physics.* Baltimore, MD: Johns Hopkins University Press, 2003.

Santos, Aaron. *Ballparking: Practical Math for Impractical Sports Questions.* Philadelphia: Running Press, 2012.

Taleb, Nassim Nicholas. *Fooled by Randomness: The Hidden Role of Chance in Life and in the Markets.* Incerto Book 1. New York: Penguin Random House, 2005.

———. *The Black Swan: The Impact of the Highly Improbable.* Incerto Book 2. New York: Penguin Random House, 2010.

———. *Antifragile: Things That Gain from Disorder.* Incerto Book 3. New York: Penguin Random House, 2014.

ABOUT THE AUTHOR

Shaun Comfort, M.D., MBA, is a Board-certified neurologist. He received a bachelor's degree in physics from Stony Brook University in New York, a Doctor of Medicine degree at the University of New Mexico, and a Master of Business Administration degree from Regis University, Colorado.

Dr. Comfort has twenty years of combined experience as a neurologist, medical reviewer at the U.S. Food and Drug Administration, and leader in biopharmaceutical and medical device companies. In addition, he has a previous career in electro-optics and computational physics in the aerospace industry. His current work has focused on clinical trial forecasting, predictive analytics, statistical decision theory, and operations research applications for healthcare, clinical trials, and pharmacovigilance. A frequent public speaker on machine learning for pharmacovigilance, Dr. Comfort has presented at conferences throughout the world. He has published widely in peer-reviewed journals, including recent publications on detecting adverse events in social media and the Modified Naranjo Causality Scale (MONARCS).

Dr. Comfort resides in Austin, Texas, with his wife, Hilary, where they enjoy hiking, kayaking, and grilling.

www.ingramcontent.com/pod-product-compliance
Lightning Source LLC
Chambersburg PA
CBHW052351220526
45465CB00003BA/1064